H. N. HUTCHINSON

EXTINCT MONSTERS

AND

CREATURES OF OTHER DAYS

A POPULAR ACCOUNT OF SOME OF THE LARGER FORMS OF ANCIENT ANIMAL LIFE

Elibron Classics
www.elibron.com

Elibron Classics series.

ISBN 1-4212-7283-0 (paperback)
ISBN 1-4212-7282-2 (hardcover)

This Elibron Classics Replica Edition is an unabridged facsimile of the edition published in 1910 by Chapman & Hall, Ltd., London.

EXTINCT MONSTERS AND CREATURES OF OTHER DAYS

"The possibilities of existence run so deeply into the extravagant that there is scarcely any conception too extraordinary for Nature to realise."—AGASSIZ.

PREFACE TO THE FIRST EDITION

BY DR. HENRY WOODWARD, F.R.S.

LATE KEEPER OF GEOLOGY, NATURAL HISTORY MUSEUM

I HAVE been requested by my friend Mr. Hutchinson, to express my opinion upon the series of drawings which have been prepared by that excellent artist of animals, Mr. Smit, for this little book entitled "Extinct Monsters."

Many of the stories told in early days, of Giants and Dragons, may have originated in the discovery of the limb-bones of the Mammoth, the Rhinoceros, or other large animals, in caves, associated with heaps of broken fragments, in which latter the ignorant peasant saw in fancy the remains of the victims devoured at the monster's repasts.

In Louis Figuier's *World before the Deluge* we are favoured with several highly sensational views of extinct monsters; whilst the pen of the late Dr. Kinns has furnished valuable information as to the "slimy" nature of their blood!

The late Mr. B. Waterhouse Hawkins (formerly a lithographic artist) was for years occupied in unauthorised restorations of various Secondary reptiles and Tertiary mammals, and about 1853 he received encouragement from Professor Owen to undertake the restorations of extinct animals which still adorn the lower grounds of the Crystal Palace at Sydenham.

But the discoveries of later years have shown that the Dicynodon and Labyrinthodon, instead of being toadlike in form, were lacertilian or salamander-like reptiles, with elongated bodies and moderately long tails; that the Iguanodon did not usually stand upon "all-fours," but more frequently sat up like some huge kangaroo with short fore limbs; that the horn on its snout was really on its wrist; that the Megalosaurus, with a more slender form of skeleton, had a somewhat similar erect attitude, and the habit, perhaps, of springing upon its prey, holding it with its powerful clawed hands, and tearing it with its formidable carnivorous teeth.

Although the Bernissart Iguanodon has been to us a complete revelation of what a Dinosaur really looked like, it is to America, and chiefly to the discoveries of Marsh, that we owe the knowledge of a whole series of new reptiles and mammals, many of which will be found illustrated within these pages.

Of long and short-tailed Pterodactyles we now know almost complete skeletons and details of their patagia or flying membranes. The discovery of the long-tailed feathered bird with teeth—the Archæopteryx, from the Oolite of Solenhofen, is another marvellous addition to our knowledge; whilst Marsh's great Hesperornis, a wingless diving bird with teeth, and his flying toothed bird, the Ichthyornis dispar, are to us equally surprising.

Certainly, both in singular forms of fossil reptilia and in early mammals, North America carries off the palm.

Of these the most remarkable are Marsh's Stegosaurus, a huge torpid reptile, with very small head and teeth, about twenty feet in length, and having a series of flattened dorsal spines, nearly a yard in height, fixed upon the median line of its back; and his Triceratops, another reptile bigger than Stegosaurus, having a huge neck-shield joined to its skull, and horns on its head and snout. Nor do the Eocene Mammals fall short of the marvellous, for in Dinoceras we find a beast with six horns, and sword-bayonet tusks, joined to a skeleton like an elephant's.

Latest amongst the marvels in modern palæontological discovery has been that made by Professor Fraas of the outline of the skin and fins in Ichthyosaurus tenuirostris, which shows it to have been a veritable shark-like reptile, with a high dorsal fin and broad fish-tail, so that "fish-lizard" is more than ever an appropriate term for these old Liassic marine reptiles.

As every palæontologist is well aware, restorations are ever liable to emendation, and that the present and latest book of extinct monsters will certainly prove no exception to the rule is beyond a doubt, but the author deserves our praise for the very boldness of his attempt, and the honesty with which he has tried to follow nature and avoid exaggeration. Every one will admire the simple and unaffected style in which the author has endeavoured to tell his story, avoiding, as far as possible, all scientific terms, so as to bring it within the intelligence of the unlearned. He has, moreover, taken infinite pains to study up his subject with care, and to consult all the literature bearing upon it. He has thus been enabled to convey accurate information in a simple and pleasing form, and to guide the artist in his difficult task with much wisdom and intelligence. That the excellence of the sketches is due to the artist, Mr. Smit, is a matter of course, and so is the blame, where criticism is legitimate; and no one is more sensible of the difficulties of the task than Mr. Smit himself.

Speaking for myself, I am *very well pleased* with the series of sketches; and I may say so with the greater ease and freedom from responsibility, as I have had very little to do with them, save in one or two trifling matters of criticism. I may venture, however, to commend them to my friends among the public at large as the happiest set of restorations that has yet appeared.

H. W.

AUTHOR'S PREFACE

NATURAL history is deservedly a popular subject. The manifestations of life in all its varied forms is a theme that has never failed to attract all who are not destitute of intelligence. From the days of the primitive cave-dwellers of Europe, who lived with mammoths and other animals now lost to the world; of the ancient Egyptians, who drew and painted on the walls of their magnificent tombs the creatures inhabiting the delta of the Nile; of the Greeks, looking out on the world with their bright and child-like curiosity, down to our own times, this old, yet ever new, theme has never failed. Never before was there such a profusion of books describing the various forms of life inhabiting the different countries of the globe, or the rivers, lakes, and seas that diversify its scenery. Popular writers have done good service in making the way plain for those who wish to acquaint themselves with the structures, habits, and histories of living animals; while for students a still greater supply of excellent manuals and text-books has been, and still continues to be, forthcoming.

But in our admiration for the present we forget the great past. How seldom do we think of that innumerable host of creatures that once trod this earth! How little in comparison has been done for *them!* Our natural-history books deal only with those that are alive now. Few popular writers have attempted to depict, as on a canvas, the great earth-drama that

has, from age to age, been enacted on the terr
which we behold the latest, but probably not the closing scenes.

When our poet wrote "All the world's a stage," he thought only of "men and women," whom he called "merely players," but the geologist sees a wider application of these words, as he reviews the drama of past life on the globe, and finds that animals, too, have had "their exits and their entrances;" nay, more, "the strange eventful history" of a human life, sketched by the master-hand, might well be chosen to illustrate the birth and growth of the tree of life, the development of which we shall trace briefly from time to time, as we proceed on our survey of the larger and more wonderful animals that flourished in bygone times.

We might even make out a "seven ages" of the world, in each of which some peculiar form of life stood out prominently, but such a scheme would be artificial.

There is a wealth of material for reconstructing the past that is simply bewildering; and yet little has been done to bring before the public the strange creatures that have perished.[1]

To the writer it is a matter of astonishment that the discoveries of Marsh, Cope, Leidy, and others in America, not to mention some important European discoveries, should have attracted so little notice in this country. In the far and wild West a host of strange reptiles and quadrupeds have been unearthed from their rocky sepulchres, often of incredibly huge

[1] Figuier's *World before the Deluge* is hardly a trustworthy book, and is often not up to date. The restorations also are misleading. Nicholson's *Life-History of the Earth* is a student's book. Messrs. Cassells' *Our Earth and its Story* deals with the whole of geology, and so is too diffusive; its ideal landscapes and restorations leave much to be desired. H. R. Knipe's *Nebula to Man* contains a large number of beautiful restorations and landscapes. See also Sir E. Ray Lankester's *Extinct Animals.*

proportions, and, in many cases, more weird and strange than the imagination could conceive; and yet the public have never heard of these discoveries, by the side of which the now well-known "lost creations" of Cuvier, Buckland, or Conybeare sink into the shade. For once, we beg leave to suggest, the hungry pressman, seeking "copy," has failed to see a good thing. Descriptions of some of "Marsh's monsters" and how they were found, might, one would think, have proved attractive to a public ever on the look-out for something new.

Professor Huxley, comparing our present knowledge of the mammals of the Tertiary era with that of 1859, states that the discoveries of Gaudry, Marsh, and Filhol are "as if zoologists were to become acquainted with a country hitherto unknown, as rich in novel forms of life as Brazil or South America once was to Europeans."

The object of this book is to describe some of the larger and more monstrous forms of the past—the lost creations of the old world; to clothe their dry bones with flesh, and suggest for them backgrounds such as are indicated by the discoveries of geology: in other words, to endeavour, by means of pen and pencil, to bring them back to life. The ordinary public cannot learn much by merely gazing at skeletons set up in museums. One longs to cover their nakedness with flesh and skin, and to see them as they were when they walked this earth.

Our present imperfect knowledge renders it difficult in some cases to construct successful restorations; but, nevertheless, the attempt is worth making: and if some who think geology a very dry subject, can be converted to a different opinion on reading these pages, we shall be well rewarded for our pains.

We venture to hope that those who will take the trouble to peruse this book, or even to look at its pictures, on which much labour and thought have been expended, will find pleasure in visiting the splendid geological collection at Cromwell Road. We have often watched visitors walking somewhat aimlessly among those relics of a former world, and wished that we could be of some service. But, if this little book should help them the better to understand what they see there, our wish will be accomplished.

Another object which the writer has kept in view is to connect the past with the present. It cannot be too strongly urged that the best commentary on the dead past is the living present. It is unfortunate that there is still too great a tendency to separate, as by a great gulf, the dead from the living, the past from the present, forms of life. The result of this is seen in our museums. Fossils have too often been left to the attention of geologists not always well acquainted with the structures of living animals. The more frequent introduction of fossil specimens side by side with modern forms of life would not only be a gain to the progress and spread of geological science, but would be a great help to students of anatomy and natural history. The tree of life is but a mutilated thing, and half its interest is gone when the dead branches are lopped off.[1]

It is, perhaps, justifiable to give to the term "monster" a somewhat extended meaning. The writer has therefore included in his menagerie of extinct animals one or two creatures which, though not of any great size, are nevertheless remarkable in various ways—such, for instance, as the winged reptiles, and anomalous birds with teeth, of later times, and others. Compared

[1] Of late years, since our first Edition in 1892, efforts have been made at the Natural History Museum, Cromwell Road, to carry out this idea to some extent by placing skeletons of living and recent animals by the side of those of extinct forms; and *vice versâ* fossils are sometimes introduced into the galleries devoted to recent forms, especially in the Reptile Gallery.

with living forms, these creatures appear to us as "monstrosities," and may well find a place in our collection.

The author wishes, in a few words, to thank those friends who have rendered him assistance in his task.

Dr. Henry Woodward, F.R.S., Keeper of Geology, Natural History Museum, has from the first taken a lively interest in this little book. He kindly helped the author with his advice on difficult matters, criticising some of the artist's preliminary sketches and suggesting improvements in the restorations. With unfailing courtesy he has ever been willing, in spite of many demands on his time, to place his knowledge at the disposal of both the author and artist; and in this way certain errors have been avoided. Besides this, he took the trouble to read through the proof-sheets, and made suggestions and corrections which have greatly improved the text. For all this welcome aid the author begs to return his sincere thanks.

To Mr. Smith Woodward, of the Natural History Museum, the author is also much indebted for his kindness in reading through the text and giving valuable information with regard to the latest discoveries.

The artist, Mr. J. Smit, notwithstanding the novelty of the subject and the difficulties of the task, has thrown himself heartily into the work of making the twenty-four restorations of extinct animals. To him, also, the author is greatly indebted, and considers himself fortunate in having secured the services of so excellent an artist.

To the publishers his thanks are due for their liberality in the matter of illustrations, and the readiness with which they have responded to suggestions.

With regard to minor illustrations the following acknowledgments are due:—

To the Palæontographical Society of Great Britain for permission to reproduce three of the illustrations in Sir Richard Owen's great work, *British Fossil Reptiles*, published in their yearly volumes, viz. Figs. 8, 9, 14.

To Messrs. Bell and Co. for the following cuts from the late Dr. Gideon A. Mantell's works: viz. Figs. 3, 35, 36, 42, 53, 67, 68, 73.

To Messrs. A. and C. Black for the following cuts from Owen's *Palæontology:* viz. Figs. 85, 99, 104.

Appendix III. contains a list of some of the works of which the writer has made use; but it would be impossible within reasonable limits to enumerate all the separate papers which have necessarily been consulted.

Models in *papier maché*, about 12 to 20 inches in length, of the following have been made under the supervision of the author, and may be obtained from Messrs. Newman (of Newman Street, Oxford Street, W.), viz. Tinoceras, Plesiosaurus, and Megatherium.

The numbers of the Figs. given are those in the present Edition (1910).

PREFACE TO SECOND EDITION

The appearance of a second edition affords the author a pleasant opportunity of thanking the reading public, and the Press, for the kind way in which his endeavour to popularise the results of modern Palæontology has been received. There seem to be fashions in all things—even in sciences; and perhaps the wonderful advances we have witnessed of late years in the physical sciences on the one hand, and in biological sciences on the other, may have tended to throw Palæontology somewhat into the shade. Let us hope that it will not remain there long.

A large number of illustrations have been added for the present edition, besides additional matter here and there in the text. Three of the plates have been redrawn. Plate IV. shows the Ichthyosaurus as interpreted by the latest discovery from Würtemberg.

Plate XXXVII., besides containing a valuable portrait of the late Sir Richard Owen, gives another drawing of the Dinornis skeleton.

April, 1893.

The numbers of the Plates are those of the present Edition (1910).

PREFACE TO NEW EDITION

PALÆONTOLOGY, like all other sciences, has made many advances since the year 1892, when the first edition of *Extinct Monsters* was published. This was followed in 1894 by *Creatures of Other Days.* The former is now out of print, and the latter not far off that state. Books, monographs, and papers have been, and still are, pouring out in a big continuous stream, in some cases setting forth results of the highest importance to Naturalists and Geologists. In the following pages an endeavour has been made to deal with some of this mass of material in a light and superficial way. The process may be compared to skimming samples of cream from many bowls of milk. For it is only the cream that the general public require. The reservoir of milk below would be too deep, dark, and unintelligible. In order to avoid making two volumes the greater part of *Creatures of Other Days* has been incorporated in the present book, and the two titles are therefore retained. A certain amount of condensation has been necessary in order to keep down the size of the book, and the chapters dealing with fossil fish, crocodiles, and "Relics of the Pleistocene Period" (*e.g.* Toxodon, Macrauchenia, Bos primigenius) have been omitted. A few creatures from *Extinct Monsters* are also reluctantly omitted.

Every endeavour has been made to illustrate the subject fully by means of fifty-five plates and a hundred and thirteen figures in the text. Mere descriptions without drawings would be of little use for our present purpose. Perhaps some young people who are fond of looking at living animals in the Zoological Gardens or in picture-books, may care to look at some of our illustrations of ancient monsters and compare them with the familiar living types which have in the course of ages descended from them by evolution.

Many of the illustrations are due to the kind co-operation of geologists and naturalists; and for this kind help the writer begs to tender his very sincere thanks. It has come in many ways, and from many lands; but especially from the United States and Russia, while English geologists have ever been most willing to offer help and advice.

It may be convenient here to indicate briefly what is new matter, and, at the same time, express our obligations to those who have contributed new illustrations.

In Chapter VII., dealing with that strange reptilian order the Anomodonts, the reader will see some new photographs of skulls and skeletons. These were very kindly sent by Professor Amalitzky, of Warsaw. They show some of the wonderful discoveries made by him in recent years. His workshop teems with bones of the Pareiasaurus and other remarkable beasts of Permian age discovered by him on the banks of the Northern Dwina, near Archangel, and only extracted from the hard nodules containing them with very great labour and patience. They used to be broken up for mending the roads! In the same geological period lived the strange reptile Naosaurus, seen in Plate XIII., which is from a photograph sent by Professor H. F. Osborn of the American Museum of Natural History.

Passing on to the Dinosaurs, we must again offer our thanks

to Professor H. F. Osborn, who supplied photographs reproduced in Plates XV., XVI. and XX. Plate XXVI. shows a new and probably better restoration of Stegosaurus, by Mr. J. Smit.

With regard to the Triceratops (the restoration of which is unusually difficult on account of the want of any single complete skeleton), a new attempt has been made by our friend Mr. J. Smit, based on the recently mounted and restored skeleton in the Natural History Museum, see Plates XXVII. and XXVIII. It was certainly a most strange beast, and we can only hope that in time *every* bone will be completely known. The now well-known Diplodocus has also been restored by the same artist from the cast in the Museum, see Plate XXII. Another comparatively new Dinosaur is the remarkable Polacanthus, seen in Fig. 60, reproduced by permission of Baron Nopcsa and Dr. Henry Woodward, F.R.S.

Some new matter and illustrations dealing with sea-serpents will be found in Chapter X.

The huge flying reptile or Pterodactyl Pteranodon, seen in Plate XXXIII., is from an illustration supplied by the Smithsonian Institution, U.S.A.

With regard to birds, some new matter has been added, and a restoration of the great Patagonian bird Phororhacos, kindly lent by Professor F. A. Lucas of the Brooklyn Museum, who also lent the picture of the Mammoth seen in Plate XLIX. Mr. W. P. Pycraft also kindly lent his own restoration of the oldest known bird, the Archæopteryx, seen in Fig. 79, for which the author is much obliged, as it seems to be the only reliable restoration yet published. The truly wonderful story of the evolution of the elephant can now be read in outline from the record of the rocks. This is certainly a result of which geologists may be proud. The drawings seen in Fig. 105 have been specially drawn for this book, and are based on the admirable restorations by Miss Alice B. Woodward, seen in Plates XLIV., XLV., XLVI. The latter have been kindly lent by Mr. H. R. Knipe, F.L.S., who published them a few years ago in his beautifully illustrated book *Nebula to Man.* From the same book the writer has also borrowed Plates XXIII. and XXVI., representing respectively Iguanodon and Stegosaurus.

As explained in the text, the fossil bones of the elephant's ancestors have been discovered chiefly by Dr. C. W. Andrews of the Natural History Museum, who has also been so fortunate as to discover the skull and large portions of the skeleton of that very strange Mammal, the Arsinoëtherium, seen in Plate XLIII. This also may be reckoned among the recent wonders of Geology.

The evolution of the horse, quite an old story now, has been brought up to date and further illustrated by the outline restorations taken from Mr. H. S. Lull, and by the Figs. on p. 245, which have been carefully drawn from casts of skulls and limb-bones in the Natural History Museum.

In the last two chapters will be found an epitome of recent information about the great Ground Sloth of South America, and the Diprotodon, a huge Marsupial from Australia.

The Trustees of the British Museum have kindly lent the following: Figs. 29, 31, 32, 37, 50, 77, 84, 99, 100, 101, 102, 103.

Dr. Henry Woodward, Editor of the *Geological Magazine,* has kindly supplied Fig. 60 of the Polacanthus, as well as Plates III., V., and LV.

In conclusion, the writer wishes to thank Dr. A. Smith Woodward, F.R.S., of the Natural History Museum, who, besides reading the proofs of this book, has from time to time offered suggestions and criticisms of sketches for which the writer is most grateful. Dr. C. W. Andrews and Mr. W. P. Pycraft have also given much help of a similar nature. The kind advice of such learned gentlemen is greatly valued by their friend the Author, for they know practically all that goes on in the world of Palæontology.

H. N. HUTCHINSON.

LONDON, *March*, 1910.

CONTENTS

LIST OF PLATES

LIST OF FIGURES IN TEXT

EXTINCT MONSTERS

INTRODUCTION

> "The earth hath gathered to her breast again
> And yet again, the millions that were born
> Of her unnumbered, unremembered tribes."

LET us see if we can get some glimpses of the primæval inhabitants of the world, that lived and died while as yet there were no men and women having authority over the fishes of the sea and the fowls of the air.

We shall, perhaps, find this antique world quite as strange as the fairy-land of Grimm or Lewis Carroll. True, it was not inhabited by "slithy toves" or "jabber-wocks," but by real beasts, of whose shapes, sizes, and habits much is already known—a good deal more than might at first be supposed. And yet, real as it all is, this antique world—this panorama of scenes that have for ever passed away—is a veritable fairy-land. In those days of which geologists tell us, the principal parts were played, not by kings and queens, but by creatures many of which were very unlike those we see around us now. And yet it is no fairy-land after all, where impossible things happen, and where impossible dragons figure largely; but only the same old world in which you and I were born. Everything you will see here is quite true. All these monsters once lived. Truth is stranger than fiction; and perhaps we shall enjoy our visit to this fairy-land all the more for that reason. For not even the dragons supposed to have been slain by armed knights in old times, when people gave ear to any tale, however extravagant, could equal in size or strength the real dragons we shall presently meet with, whose actual bones may be seen in the Natural History Museum at South Kensington.

Many people who visit this great museum and find their way to the geological galleries on the right, pass hastily by the cases of bones, teeth, and skeletons. These things, it seems, fail to interest them. They do not know how to interpret them. They cannot picture to themselves the kinds of creatures to which the relics once belonged; and so they pass them by and presently go to the more attractive collection of stuffed birds on the other side. There they see the feathered tribes of the air all beautifully arranged; some poised in the air by almost invisible wires; some perched on branches: but all surrounded by grass, flowers, and natural objects, imitated with marvellous reality, so that they see the birds as they really are in nature, and can almost fancy they hear them singing.

Now, it has often occurred to the present writer that something more might be done for the long-neglected "lost creations" of the world, to bring them out of their obscurity, that they may be made to tell to the passer-by their wondrous story. We can, however, well imagine some of our readers asking, "Can these dry bones live?" "Yes," we would say, "they can be made to live; reason and imagination will, if we give them proper play, provide us eyes wherewith to see the world's lost creations." To such men as Cuvier, Owen, Huxley, and others, these dry bones *do* live. It will be our object to describe to the reader some of the wonderful results that have rewarded the lifelong labours of such great men. We shall take some of the largest and strangest forms of life that once lived, and try to picture them as they really were when alive, whether walking on land, swimming in the sea, or flying in the air; to understand the meanings of their more obvious structures; and to form some conclusions with regard to their habits, as well as to find out, if possible, their relations,—as far as such questions have been answered by those most qualified to settle these difficult matters.

All technical details, such as the general reader is unfamiliar with, will be as far as possible suppressed. Let us fancy a long procession of extinct monsters passing in single file before us, and ourselves endeavouring to pick out their "points" as they present themselves to the eye of imagination. It is not, be it remembered, mere imagination that guides the man of science in such matters, for all his conclusions are carefully based on reason; and when conclusions are given, we shall endeavour to show how they have been arrived at.

For millions of years countless multitudes of living animals have played their little parts on the earth and passed away, to be buried up in the oozy beds of the seas of old time, or entombed with the leaves that sank in the waters of primæval lakes. The majority of these perished beyond all recovery, leaving not a trace behind; yet a vast number of fossilised remains have been, in various ways, preserved; sometimes almost as completely as if Dame Nature had thoughtfully embalmed them for our instruction and delight.

Down in those old seas and lakes she kept her great museum, in order to preserve for us a selection of her treasures. In course of time she slowly raised up sea-beds and lake-bottoms to make them into dry land. This museum is everywhere around us. We have but to enter quarries and railway cuttings, or to search in coal-mines, or under cliffs at the seaside, and we can consult her records. As the ancient Egyptians built tombs, pyramids, and temples, from which we may learn their manner of life and partly read their history, so Nature has entombed, not one race only, but many races of the children of life. Her records are written in strange hieroglyphs, yet it is not difficult to interpret their meaning; and thus many an old story, many an old scene, may be pictured in the mind of man.

Shall we call this earth-drama a tragedy or a comedy? Doubtless tragic scenes occurred at times; as, for instance, when fierce creatures engaged in deadly combat: and probably amusing, if not comic, incidents took place occasionally, such as might have provoked us to laughter, had we been there to see them. But let us simply call it a drama. Backgrounds of scenery were not wanting. Then, as now, the surface of the earth was clothed with vegetation, and strange cattle pastured on grassy plains. Vegetation was at times very luxuriant. The forests of the coal period, with their giant reeds and club-moss trees, must have made a strange picture. Then, as now, there rose up from the plains lofty ranges of mountains, reaching to the clouds, their summits clothed with the eternal snows. These, too, played their part, feeding the streams and the rivers that meandered over the plains, bringing life and fertility with them, as they do now. The sun shone and the wind blew: sometimes gently, so that the leaves just whispered in an evening breeze; at other times so violently that the giants of the forest swayed to and fro, and the seas lashed themselves furiously against rocky coasts. Nor were the underground forces of the earth less active than they are now: volcanic eruptions often took place on a magnificent scale; volcanoes poured out fiery lava-streams for leagues beneath their feet; great showers of ashes and fine dust were ejected in the air, so that the sun was darkened for a time, and the surface of the sea was covered for many miles with floating pumice and volcanic dust, which in time sank to the bottom, and was made into hard rock, such as we now find on the top of Snowdon.

Earthquake shocks were quite as frequent, and no doubt the ground swayed to and fro, or was rent open as some unusually great earth-movement took place, and perhaps a mountain range was raised several feet or yards higher. All this we learn from the testimony of the rocks beneath our feet. It only requires the use of a little imagination to conjure up scenes of the past, and paint them as on a moving diorama.

We shall not, however, dwell at any length on the scenery, or the vegetation that clothed the landscape at different periods; for these features are sufficiently indicated in the beautiful drawings of extinct animals by our artists, Mr. J. Smit, and others.

The researches of the illustrious Baron Cuvier, at Paris, as embodied in his great work, *Ossemens Fossiles*, gave a great impetus to the study of organic remains. It was he who laid the foundations of the science of Palæontology,[1] which, though much has already been accomplished, yet has a great future before it. Agassiz, Owen, Huxley, Marsh, Cope, and others, following in his footsteps, have greatly extended its boundaries; but he was the pioneer.

Before Cuvier's time fossil forms were very little known, and still less understood. His researches, especially among vertebrates, or backboned animals, revealed an altogether undreamed-of wealth of entombed remains. It is true, the old and absurd notion that

[1] Palæontology is the science which treats of the living beings, whether animal or vegetable, which have inhabited this globe at past periods in its history. (Greek—*palaios*, ancient; *onta*, beings; *logos*, discourse.)

fossils were mere "sports of Nature," sometimes bearing more or less resemblance to living animals, but still only an accidental (!) resemblance, had been abandoned by Leibnitz, Buffon, and Pallas; and that Daubenton had actually compared the fossil bones of quadrupeds with those of living forms; while Camper declared his opinion that some of these remains belonged to extinct species of quadrupeds.

It is to Cuvier, however, that the world owes the first systematic application of the science of comparative anatomy, which he himself had done so much to place on a sound basis, to the study of the bones of fossil animals. He paid great attention to the relative shapes of animals, and the different developments of the same kind of bones in various animals, and especially to the nature of their teeth. So great did his experience and knowledge become, that he rarely failed in naming an animal from a part of its skeleton. He appreciated more clearly than others before him the mutual dependence of the various parts of an animal's organisation. "The organism," he said, "forms a connected unity, in which the single parts cannot change without modifications in the other parts."

It will hardly be necessary to give examples of this now well-known truth; but, just to take one case: the elephant has a long proboscis with which it can reach the ground, and consequently its neck is quite short; but take away the long proboscis, and you would seriously interfere with the relation of various parts of its structure to each other. How, then, could it reach or pick up anything lying on the ground? Other changes would have to follow: either its legs would require to be shortened, or its neck to be lengthened. In every animal, as in a complex machine, there is a mutual dependence of the different parts.

As he progressed in these studies, Cuvier was able with considerable success to restore extinct animals from their fossilised remains, to discover their habits and manner of life, and to point out their nearest living ally. To him we owe the first complete demonstration of the possibility of restoring an extinct animal. His "Law of Correlation" however, has been found to be not infallible; as Professor Huxley has shown, it has exceptions. It expresses our experience among living animals, but, when applied to the more ancient types of life, is liable to be misleading.

To take one out of many examples of this law: Carnivorous animals, such as cats, lions, and tigers, have claws in their feet, very different from the hoofs of an ox, which is herbivorous: while the teeth of the former group are very different from those of the latter. Thus the teeth and limbs have a certain definite relation to each other, or, in other words, are correlated. Again, horned quadrupeds are all herbivorous (or graminivorous), and have hoofs to their feet. The following amusing anecdote serves to illustrate Cuvier's law. One of his students thought he would try and frighten his master, and, having dressed up as a wild beast, entered Cuvier's bedroom by night, and, presenting himself by his bedside, said in hollow tones, "Cuvier, Cuvier, I've come to eat you!" The great naturalist, who on waking up was able to discern something with horns and hoofs, simply remarked, "What! horns, hoofs—graminivorous—you can't!" What better lesson could the master have given the pupil to help him to remember his "Law of Correlation"?

Cuvier's great work, entitled *Ossemens Fossiles*, will long remain an imperishable monument of the genius and industry of the greatest pioneer in this region of investigation. This work proved beyond a doubt to his astonished contemporaries the great antiquity of the tribes of animals now living on the surface of the earth. It proved more than that, however; for it showed the existence of a great philosophy in Nature which linked the past with the present in a scheme that pointed to a continuity of life during untold previous ages. All this was directly at variance with the prevalent ideas of his time, and consequently his views were regarded by many with alarm, and he received a good deal of abuse—a fate which many other original thinkers before him have shared.

It is somewhat difficult for people living now, and accustomed to modern teaching, to realise how novel were the conclusions announced by Cuvier. In his *Discourse on the Revolutions of the Surface of the Globe*, translated into most European languages under the title *Theory of the Earth*, he lays down, among others, the two following propositions:—

1. That all organised existences were not created at the same time; but at different times, probably very remote from each other—vegetables before animals, mollusca and fishes before reptiles, and the latter before mammals.

2. That fossil remains in the more recent strata are those which approach nearest to the present type of corresponding living species.

Teaching such as this gave a new impetus to the study of organic remains, and Palæontology, as a science, began with Cuvier.

The reader is recommended to study Chapters X. and XI. of Darwin's *Origin of Species*, dealing with the imperfection of the record, distribution, and extinction of species, etc.

One object which the writer has kept in view all through is to show how strong is the evidence for Evolution. In fact, it is overwhelming; consequently in these days one seldom meets a naturalist or a geologist who does not accept the idea of Evolution. But when people speak of "Darwinism" they do not always clearly distinguish in their minds between Evolution and the means by which evolutionary changes have been brought about. Darwin tried to prove that the great agency (so long sought for in vain) was Natural Selection. Hence some people mean Natural Selection when they speak of Darwinism. Since Darwin's death in the year 1882 a great many naturalists and others have pondered deeply on this great subject; new facts and new conceptions have arisen on the question of the Origin of Species. Some of our readers may perhaps think that Darwin settled the question once for all. But that is not so. True, there is a powerful school of naturalists (Neo-Darwinians) who never weary of telling us that Natural Selection is the great and principal cause, or agency, whereby new species arose. On the other hand, there are many living naturalists of renown who are unwilling to accept Darwin's explanation. Moreover, to some minds such an explanation is far too mechanical, involving a narrow view of the beauties and wonders of Creation, as if Utility were the principal thing. Again, it is possible that the struggle between individuals of the same or near species is not so great as the Darwinian conception demanded. Darwin's argument was largely based on the analogy of the sort of selection used by breeders of plants and domestic animals. It now appears that this analogy is not quite safe. Some naturalists believe that the great progress Man has made in breeding pigeons, cattle, horses, dogs, and flowers may all be based on definite and sudden changes, now known as "Mutations" (sports). Mendel's Law was unknown to Darwin, and the striking suggestions of De Vries had not appeared in his day. In conclusion, it only remains to say (for our space is limited) that at present the great problem is unsolved. Perhaps the time is hardly ripe.

CHAPTER I

HOW EXTINCT MONSTERS ARE PRESERVED

"Geology, beyond almost every other science, offers fields of research adapted to all capacities and to every condition and circumstance of life in which we may be placed. For while some of its phenomena require the highest intellectual powers and the greatest attainments in abstract science for their successful investigation, many of its problems may be solved by the most ordinary intellect, and facts replete with the deepest interest may be gleaned by the most casual observer."—MANTELL.

LET us suppose we are visiting a geological museum for the first time, passing along from one department to another with ever-increasing wonder—now admiring the beautiful polished marbles from Devonshire, with their delicate corals, or the wonderful fishes from the Old Red Sandstone, with their plates of enamel; now the delicate shells and ammonites from the Lias or Oolites, with their pearly lustre still preserved; now the white fresh-looking shells from the Isle of Wight; now the ponderous bones and big teeth of ancient monsters from the Wealden beds of Sussex. The question might naturally occur, "How were all these creatures preserved from destruction and decay, and sealed up so securely that it is difficult to believe they are as old as the geologists tell us they are?" It will be worth our while to consider this before we pass on to describe the creatures themselves.

Now, in the first place, "fossils" are not always "petrifactions," as some people seem to think; that is to say, they are not all turned into stone. This is true in many cases, no doubt, yet one frequently comes across the remains of plants and animals that have undergone very little change, and have, as it were, been simply sealed up. The state of a fossil depends on several circumstances, such as the soil, mud, or other medium in which it may happen to be preserved. Again, the newest, or most recent, fossils are generally the least altered. We have fossils of all ages, and in all states of preservation. As examples of fossils very little altered, we may take the case of the wonderful collection of bones discovered by Professor Boyd Dawkins in caves in various parts of Great Britain. The results of many years of research are given in his most interesting book on *Cave-Hunting*. This enthusiastic explorer and geologist has discovered the remains of a great many animals, some of which are quite extinct, while others are still living in this country. These remains belong to a late period, when lions, tigers, cave-bears, wolves, hyænas, and reindeer inhabited our country. In some cases the caves were the dens of hyænas, who brought their prey into caverns in our limestone rocks, to devour them at their leisure; for the marks of their teeth may yet be seen on the bones. In other cases the bones seem to have been washed into the caves by old streams that have ceased to run; but in all cases they are fairly fresh, though often stained by iron-rust brought in by water that has dissolved iron out of various rocks—for iron is a substance met with almost everywhere in nature. Sometimes they are buried up in a layer of soil, or "cave-earth," and at other times in a layer of stalagmite—a deposit of carbonate of lime gradually formed on the floors of caves by the evaporation of water charged with carbonate of lime.

Air and water are great destroyers of animal and vegetable substances from which life has departed. The autumn leaves that fall by the wayside soon undergo change, and become at last separated or resolved into their original elements. In the same way when any wild animal, such as a bird or rabbit, dies in an exposed place, its flesh decays under the influence of rain and wind, so that before long nothing but dry bones is left. Hamlet's wish that this "too too solid flesh would melt" is soon realised after death; and that active chemical element in the air known as oxygen, in breathing which we live, has a tenfold power over dead matter, slowly causing chemical actions somewhat similar to those that take place in a burning candle, whereby decaying flesh is converted into water-vapour and carbonic acid gas. Thus we see that oxygen not only supports life, but breaks up into simpler forms the unwholesome and dangerous products of decaying matter, thus keeping the atmosphere sweet and pure; but in time, even the dry bones of the bird or rabbit, though able for a longer period to resist the attacks of the atmosphere, crumble into dust, and serve to fertilise the soil that once supported them.

Now, if water and air be excluded, it is wonderful how long even the most perishable things may be preserved from this otherwise universal decay. In the Edinburgh museum of antiquities may be seen an old wooden cask of butter that has lain for centuries in peat—which substance has a curiously preservative power; and human bodies have been dug out of Irish peat with the flesh well preserved, which, from the nature of the costume worn by the person, we can tell to be very ancient. Meat packed in tins, so as to be entirely excluded from the air, may be kept a very long time, and will be found to be quite fresh and fit for use.

But air and water have a way of penetrating into all sorts of places, so that in nature they are almost everywhere. Water can slowly filter through even the hardest rocks, and since it contains dissolved air, it causes the decay of animal or vegetable substances. Take the case of a dead leaf falling into a lake, or some quiet pool in a river. It sinks to the bottom, and is buried up in gravel, mud, or sand. Now, our leaf will stand a very poor chance of preservation on a sandy or gravelly bottom, because these materials, being porous, allow the water to pass through them easily. But if it settles down on fine mud it may be covered up and become a fossil. In time the soft mud will harden into clay or shale, retaining a delicate impression of the leaf; and even after thousands of years, the brown body of the leaf will be there only partly changed. In the case of the plants found in coal, the lapse of ages since they were buried up has been so great (and the strata have been so affected by the great pressure and by the earth's internal heat) that certain chemical changes have converted leaves and stems into carbon and some of its compounds, much in the same way that, if you heat wood in a closed vessel, you convert it into charcoal, which is mostly carbon. The coal we burn in our fires is entirely of vegetable origin, and every seam in a coal-mine is a buried forest of trees, ferns, reeds, and other plants.

The reader will understand how it is that rocks composed of hardened sand or gravel, sandstones and conglomerates, contain but few fossils; while, on the other hand, such rocks as clay, shale, slate, and limestone often abound in fossils, because they are formed of what was once soft mud, that sealed up and protected corals, shell-fish, sea-urchins, fishes, and other marine animals. Had they been covered up in sand the chances are that percolating water would have slowly dissolved the shells and corals, the hard coats of the crabs, and the bones of the fishes, all of which are composed of carbonate of lime; and we know that is a substance easily dissolved by water.

It is in the rocks formed during the later geological periods that we find fossils least changed from their original state; for time works great changes, and too little time has elapsed since those periods for any considerable alterations to have taken place. But when we come to examine some of the earlier rocks, which have been acted upon in various ways for long periods of time, such as the pressure of vast piles of overlying rocks, and the percolation of water charged with mineral substances (water sometimes warmed by the earth's internal heat), then we may expect to find the remains of the world's lost creations in a much more mineralised condition. Every fossil-collector must be familiar with examples of changes of this kind. For instance, shells originally composed of carbonate of lime are often found to have been turned into flint or silica. Another curious change is illustrated in the case of a stratum found in Cambridgeshire and other counties. In this remarkable layer, only about a foot in thickness, one frequently finds bones and teeth of fishes and reptiles. These, however, have all undergone a curious change, whereby they have been converted into phosphate of lime—a compound of phosphorus and lime. It abounds in "nodules," or lumps, of this substance, which along with thousands of fossils, are every year ground up and converted by a chemical process into valuable artificial manure for the farmer.

The soft parts of animals, as we have said before, cannot be preserved in a fossil state; but, as if to compensate for this loss, we sometimes meet with the most faithful and delicate impressions. Thus, cuttle-fishes have, in some instances, left, on the clays which buried them up, impressions of their soft, long arms, or tentacles, and, as the mud hardened into solid rock, the impressions are fixed imperishably. Examples of these interesting records may be seen at the Natural History Museum at South

Kensington. Even soft jelly-fishes have left their mark on certain rocks! At a place in Bavaria, called Solenhofen, there is a remarkably fine-grained limestone containing a multitude of wonderful impressions. This stone is well known to lithographers, and is largely used in printing. On it the oldest known bird has left its skeleton and faithful impressions of its feathers.

The footprints of birds and reptiles are by no means uncommon. Such records are most valuable, for a great deal may be learned from even a footprint as to the nature of the animal that made it (see p. 41).

Since the greater number of animals described in this book are reptiles, quadrupeds, and other inhabitants of the land, and only a few had their home in the sea, we must endeavour to try and understand how their remains may have been preserved. Our object in writing this book is to interpret their story, and, as it were, to bring them to life again. Each one must be made to tell its own story, and that story will be far from complete if we cannot form some idea of how it found its way into a watery grave, and so was added to Nature's museum. For this purpose we must briefly explain to the reader how the rocks we see around us have been deposited; for these rocks are the tombs in which lost creations lie.

Go into any ordinary quarry, where the men are at work, getting out the stone in blocks to be used in building, or for use on the roads, or for some other purpose, and you will be pretty sure to note at the first glance that the rock is arranged as if it had been built up in layers. Now, this is true of all rocks that have been laid down by the agency of water—as most of them have been. True, there are exceptions, but every rule has its exceptions. If you went into a granite quarry at Aberdeen, or a basalt quarry near Edinburgh, you would not see these layers;

but such rocks as these do not contain fossils. They have been mainly formed by the action of great heat, and were forced up to the surface of the earth by pressure from below. As they slowly cooled, the mineral substances of which they were formed gradually crystallised; and it is this crystalline state, together with the signs of movement, that tells us of their once heated state. Such rocks are said to be of igneous origin (Lat. *ignis*, fire). But nearly all the other rocks were formed by the action of water—that is, under water,—and hence are known to geologists as aqueous deposits (Lat. *aqua*, water). They may be considered as sediments that slowly settled down in seas, lakes, or at the mouths of rivers. Such deposits are in the course of being formed at the present day. All round our coasts mud, sand, and gravel are being accumulated, layer by layer. These materials are constantly being swept off the land by the action of rain and rivers, and carried down to the sea. Perhaps, when staying at the seaside, you may have noticed, after rainy and rough weather, how the sea, for some distance from the shore, is discoloured with mud, especially at the mouth of a river. The sand, being heavy, soon sinks down, and this is the reason why sand-bars so frequently block the entrance to rivers. Then, again, the waves of the sea beat against the seashore and undermine the cliffs, bringing down great fragments, which after a time are completely broken up and worn down into rounded pebbles, or even fine sand and mud. It is very easy to see that in this way large quantities of sand, gravel, and mud are continually supplied to our seas. We can picture how they will settle down; the sand not far from the shore, and the fine mud further out to sea. When the rough weather ceases, the river becomes smaller and flows less rapidly, so that when the coarse *débris* of the land has settled down to form layers, or strata, of sand and gravel, then the fine mud will begin to settle down also, and will form a layer overlying them or further out. Then we learn, from a little observation of what is now going on, how layers of sand and mud, such as we see in a quarry, were made thousands and thousands of years ago.

When we think of all the big rivers and small streams continually flowing into the sea, we shall begin to realise what a great work rain and rivers are doing in making the rocks of the future. If, at a later period, a slight upheaval of the sea-bed were to take place so as to bring it above water—and such is very likely—these materials would be found neatly arranged in layers, and more or less hardened into solid rock.

The reader may, perhaps, find it rather hard at first to realise that in this simple way vast deposits of rock are being formed in the seas of the present day, and that the finer material thus derived from a continent may be carried by ocean currents to great distances; but so it is. Over thousands of square miles of ocean, deposits are being gradually accumulated which will doubtless be some day turned into hard rock. Just to take one example: it has been found that in the Atlantic Ocean, at a distance of over two hundred miles from the mouth of that great river, the Amazon, the sea is discoloured by fine sediment.

There is another kind of rock frequently met with, the building up of which cannot be explained in the way we have pointed out, and that is limestone. This rock has not been deposited as a sediment, like clays and sandstones, but geologists have good reasons for believing that it has been gradually formed in the deeper and clearer parts of oceans by the slow accumulation of marine shells, corals, and other creatures, whose bodies are partly composed of carbonate of lime. This seems incredible at first, but the proofs are quite convincing.[1] As Professor Huxley well remarked, there is as good evidence that chalk has been built up by the accumulation of minute shells as that the Pyramids were built by the ancient Egyptians.

The science of geology reveals the startling fact that all the great series of the stratified rocks, whose united thickness is over 80,000 feet, has been mainly accumulated under water, either by the action of those powerful geological agents—rain and rivers—or through the agency of myriads of tiny marine animals. When we have grasped this idea, we have learned our first, and perhaps most useful, lesson in geology.

Now let us apply what has been above explained to the question immediately before us. We want to know how the skeletons of animals living on land came to be buried up under water, among the stratified rocks that are to be seen all over our country, and most of which were made under the sea.

We can answer this question by going to Nature herself, in order to find out what is actually going on at the present time, by inquiring into the habits of land animals, their surroundings, and the accidents to which they are liable at sundry times and in divers manners. It is by this simple method of studying present actions that nearly all difficult questions in geology may be solved. The leading principle of the geologist is to interpret the past by the light of the present; or, in other words, to find out what happens now, in order to learn what took place ages ago; for it is clear that the world has been going on in the same way for at least as far back as geological history can take us. There has been a *uniformity*, or sameness, in Nature's actions ever since living things first dwelt on the earth.

Just as rivers are mainly responsible for bringing down to the

[1] See the writer's *Autobiography of the Earth*, p. 223.

sea the materials of which rocks are made, so these universal carrying agents are the means by which the bodies of many animals that live in the plains, over which they wander, are brought to their last resting-place. We have only to consult the records of great floods to see what fearful havoc they sometimes make among living things, and how the dead bodies are swept away.

Great floods rise rapidly, so that the herds of wild animals pasturing on grassy plains are surprised by the rising waters, and, being unable to withstand the force of the water, are hurried along, and so drowned. When dead they sink to the bottom, and may, in some cases, be buried up in the *débris* hurried along by the river; but as a rule their bodies, being swollen by the gases formed by decomposing flesh, rise again to the surface, and consequently may be carried along for many a mile, till they reach some lake, or perhaps right down to the mouth of a river, and so may be taken out to sea.

One or two examples will be given to show how important is the action of such floods. Sir Charles Lyell has given some striking illustrations of this. There was a memorable flood in the southern borders of Scotland on the 24th of June, 1794, which caused great destruction in the region of the Solway Firth. Heavy rains had fallen, so that every stream entering the firth was greatly swollen. Not only sheep and cattle, but even herdsmen and shepherds were drowned. When the flood had subsided, a fearful spectacle was seen on a large sandbank, called "the beds of Esk," where the waters meet; for on this one bank were found collected together the bodies of 9 black cattle, 3 horses, 1840 sheep, 45 dogs, 180 hares, together with those of many smaller animals, also the corpses of two men and one woman.

Humboldt, the celebrated traveller, says that when, at certain seasons, the large rivers of South America are swollen by heavy rains, great numbers of quadrupeds are drowned every year. Troops of wild horses that graze in the "savannahs," or grassy plains, are said to be swept away in thousands.

In Java, in the year 1699, the Batavian River was flooded during an earthquake, and drowned buffaloes, tigers, rhinoceroses, deer, apes, crocodiles, and other wild beasts, which were brought down to the coast by the current.

In tropical countries, where very heavy rains fall at times, and rivers become rapidly swollen, floods are a great source of danger to man and beast. Probably the greater number of the bodies of animals thus drowned find their way into lakes, through which rivers flow, and never reach the sea; and if the growth of sediment in such lakes goes on fairly rapidly, their remains may be buried up, and so preserved. But in many cases the bones fall one by one from the floating carcase, and so may in that way be scattered at random over the bottom of the lake, or the bed of a river at its mouth. In hot countries such bodies, on reaching the sea, run a great chance of being instantly devoured by sharks, alligators, and other carnivorous animals. But during very heavy floods, the waters that reach the sea are so heavily laden with mud, that these predaceous animals are obliged to retire to some place where the waters are clear, so that at such times the dead bodies are more likely to escape their ravages; and, at the same time, the mud with which the waters are charged falls so rapidly that it may quickly cover them up. We shall find further on that this explanation probably applies to the case of the "fish-lizards," whose remains are found in the Lias formation (see p. 75).

But, for several reasons, sedimentary rocks formed in lakes are much more likely to contain the remains of land animals than those that were formed in seas, and they are more likely to be in a complete state of preservation. Within the last century, five or six small lakes in Scotland, which had been artificially drained, yielded the remains of several hundred skeletons of stags, oxen, boars, horses, sheep, dogs, hares, foxes, and wolves. There are two ways in which these animals may have met with a watery grave. In the first place, they may have got mired on going into the water, or in trying to land on the other side, after swimming across. Any one who knows Scottish lakes will be familiar with the fact that their margins are often most treacherous ground for bathers. The writer has more than once found it necessary to be very cautious on wading into a lake while fishing, or in search of plants. Secondly, when such lakes are frozen over in winter, the ice is often very treacherous in consequence of numerous springs; and animals attempting to cross may be easily drowned. No remains of birds were discovered in these lakes, in spite of the fact that, until drained, they were largely frequented by water-fowl. But it must be remembered that birds are protected by their powers of flight from perishing in such ways as other animals frequently do. And, even should they die on the water, their bodies are not likely to be submerged; for, being light and feathery, they do not sink, but continue floating until the body rots away, or is devoured by some creature such as a hungry pike. For these reasons the remains of birds are unfortunately very rare in the stratified rocks; and hence our knowledge of the bird life of former ages is slight.

The Imperfection of the Record.

A very little consideration will serve to convince us that the record which Nature has kept in the stratified rocks is an incomplete one. There are many reasons why it must be so. It is not to be expected that these rocks should contain anything like a complete collection of the remains of the various tribes of plants and animals that from time to time have flourished in seas, lakes, and estuaries, or on islands and continents of the world. In endeavouring to trace the course of life on the globe at successive periods, we are continually met by want of evidence due to the "imperfection of the record"—to use Darwin's phrase. The reasons are not far to seek. The preservation of organic remains, or even of impressions thereof, in sedimentary strata is, to some extent, a matter of chance. It is obvious that no wholly soft creature, such as a jelly-fish, can be preserved; although on some strata they have left impressions telling of their existence at a very early period.

A creature, to become fossilised, must possess some hard part, such as a shell, *e.g.* an oyster (fossil oysters abound in some strata); or a hard chitinous covering, like that of the shrimp, or the trilobites of Silurian times; or a skeleton, such as all the backboned (vertebrate) animals possess.

But even creatures that had skeletons have not by any means always been preserved. Bones, when left on the bottom of the sea, where no sediment, or very little, is forming, will decay, and so disappear altogether. As Darwin points out, we are in error in supposing that over the greater part of the ocean-bed of the present day sediment is deposited fast enough to seal up organic remains before they can decay. Over a large part of the ocean-bed such cannot be the case; and this conclusion has, of late years, been confirmed by the observations made during the fruitful voyage of H.M.S. *Challenger* in the Atlantic and Pacific Oceans.

Again, even in shallower parts of the old seas, where sand or mud was once deposited fossilisation was somewhat accidental; for some materials, being porous, allow of the percolation of water, and in this way shells, bones, etc., have been dissolved and lost. Thus sandstone strata are nearly barren in fossils compared to shales and limestones, which are much less pervious. To take examples from our own country, the New Red Sandstone of the south-west of England, the midland counties, Cheshire, and other parts contains very few fossils indeed, while the clays and limestones of the succeeding Lias period abound in organic remains of all sorts. Even insects have left delicate impressions of their wings and bodies! while shells, corals, encrinites, fish-teeth, and bones of saurians are found in great numbers.

Again, it must be borne in mind that the series of stratified rocks known to geologists is not complete or unbroken. They have been well compared to the leaves of a book on history, of which whole chapters and many separate pages have been torn out. These gaps, or "breaks," are due to what is called "denudation;" that is to say, a great many rocks, after having been slowly deposited in water, have been upraised to form dry land, and then, being subjected for ages to the destroying action of "rain and rivers," or the waves of the sea, have been largely destroyed. Such rocks, in the language of geology, have been "denuded;" that is, stripped off, so that the underlying rocks are left bare.

But the process of rock-making does not go on continuously in any one area. Sedimentary strata have been formed in slowly sinking areas. But, if subsidence ceases, and the downward movement becomes an upward one, then the bed of the sea is converted into dry land, and the geological record is broken; for aqueous strata do not form on dry land. Blown sands and terrestrial lava-flows are exceptions; but such accumulations are very small and insignificant, and may therefore be neglected, especially as they contain no fossils.

In this way, as well as by the process of "denudation" already alluded to, breaks occur; and these breaks often represent long intervals of time. There are several such gaps in the British series of stratified rocks; and it is partly by means of these breaks, during which important geographical and other changes took place, that sedimentary rocks have been classified and arranged in groups representing geological periods. Thus, the Cainozoic, or Tertiary, rocks of the Thames basin are separated by a long "break" from those of the preceding Cretaceous period. During that interval great changes in animal life took place, whereby, in the course of evolution, new types appeared on the scene. (See Table of Strata, Appendix I.)

Another cause interfering with the record is to be found in those important internal changes that have taken place in stratified rocks—often over large areas—which may be ascribed to the influence of heat and pressure combined. This process of change, whereby soft deposits have been altered or "metamorphosed" into hard crystalline rocks, is known as "metamorphism." Metamorphic rocks have lost not only their original structure and appearance, but also their included organic remains, or fossils. Thus, when a soft limestone has been converted by these means into crystalline statuary marble, any fossils it may once have contained have been destroyed. It is true that this applies more to older and lower deposits—for the lowest are the oldest—but there can be no doubt that valuable records of the forms of life which peopled the world in former periods have been lost by this means.

And lastly, it must ever be borne in mind that, as yet, our knowledge of the stratified rocks of the earth's crust is very limited. In course of time, no doubt, this deficiency will be to a great extent made good; but it will take a long time. Already, within the last thirty years, the labours of zealous geologists in the colonies and in various countries have added largely to our knowledge of the geological record. Still, only a small portion of the earth's surface has at present been explored; and doubtless one may look forward to future discoveries of extinct forms of animal and plant life as wonderful and strange as those that have been of late years unearthed in the "far West," in Africa, and India. The Siwalik Hills of Northern India offer a rich harvest of fossils to future explorers. Already one remarkable and large horned quadruped has come from this region; and it is known that other valuable treasures are sealed up within these hills, only awaiting the "open sesame" of some enterprising explorer to bring them to light.

As previously pointed out, deposits formed in lakes are the most promising field for geologists in search of the remains of old terrestrial quadrupeds and reptiles; but, unfortunately, such deposits are rare.

It is very much to be regretted that the carelessness and indifference of ignorant workmen in quarries, clay-pits, and railway cuttings have sometimes been the cause of valuable fossils being broken up, and so lost for ever. Unless they are accustomed to the visits of fossil-collectors who will pay them liberally for their finds, the men will not take the trouble to preserve any bones they may come across in the course of their work. (Examples of this negligence will be found on pp. 41, 65, and 295.) But when once they realise that such finds have what political economists call an "exchange value," or, in other words, can be turned into money, it is astonishing what zealous guardians of Nature's treasures they become! For this reason collectors often find what Professor Bonney calls the "silver hammer"—in other words, cash—more effective than the steel implement they carry with them.

CHAPTER II

FOOTPRINTS ON THE SANDS OF TIME

"They are fraught with strange meanings, these footprints of Connecticut." —Hugh Miller.

There is a great deal of truth in the saying of Emerson, that "everything in Nature is engaged in writing its own history." The more one studies the changes taking place every day on the surface of the earth, in order to read the riddle of the rocks beneath our feet, which contain Mother Earth's records of her past history, the more one is impressed with the truth of this saying. In fact, it is not too much to say that the whole science of Geology is founded on this idea. The geologist is he who interprets to his fellows the stony documents contained in Nature's "Record Office," and he finds the key to the interpretation of her hieroglyphics in watching her daily actions at the present time. One branch of this science, viz. Physical Geology, deals with the earth's physical features, interpreting in the light of this leading principle their history, and telling us how the river carved out its valley; how the volcano was built up; how the mighty mountains were raised up from the beds of ancient seas, to be carved out by the agents of denudation into all their varied and wonderful features.

Another branch of geology, namely Palæontology, with which we are about to deal in the present work, tells us of the long-lost tribes of plants and animals which, ages and ages ago, found a home on the earth. To many minds this branch of geology, which is simply the natural history of the past, is the most fascinating, with which view we fully sympathise. Putting aside the study of fossil plants (which is a small branch of the subject), we may say of Palæontology, that it interprets to us the world's "lost creations." In this branch of geology the records are not so much the rocks themselves as the fossil skeletons they so often contain.

In some cases, as we shall presently show, the rocks contain additional evidences of very considerable value, as throwing light on the habits of the creatures, or on their natural surroundings; but bones, shells, and other hard parts of animals, are the foundation on which the science of Palæontology is founded. And here, again, we find the same principle at work—viz. that the past must be read in the light of the present. However many ages ago it it was, whether millions or billions of years ago, that these primæval inhabitants of the world enjoyed their existence, the same unbroken laws of nature—the visible expressions of a Divine and All-powerful Will—were at work, fulfilling His purposes as now. Flesh and blood were then what they are now, and fulfilled the same functions. Bones grew then as they grow nowadays. To those bones were attached muscles which expanded and contracted just as muscles do now. Wings were used for flying, fins and paddles for swimming, legs for walking, teeth for masticating food, just as they are now. In fact these primitive inhabitants of the antique world, however different in bodily shape from those we see around us now, lived under the same universal laws of Physiology as we ourselves do.

Palæontology, then, is the science which, in the light of Comparative Anatomy and Physiology, rehabilitates the world's ancient inhabitants, clothing their dry bones with flesh, and enabling us in imagination to see them as they were when they walked this earth. It will be our endeavour in the present work to present to our readers a certain number of these antique animals —birds, beasts, and fishes, whose mortal remains have been buried up and preserved, often with singular completeness, in the rocks of the earth's crust.

But although fossil bones and skeletons are the chief material at our command for this purpose, yet the series of stratified rocks contains here and there other kinds of evidence, valuable in their way, such as footmarks, tracks, burrows, coprolites, or droppings, and even ripple-marks and the impressions of rain-drops. It is with these evidences that we propose to deal in the present chapter. Let us see what can be learned from such humble and apparently insignificant records, of some of the creatures that once trod this earth.

The intelligent observer who has strolled along the strand of the seashore at low water, must have often seen the surface of the exposed sands deeply rippled by the waves of the ebbing tide, and have noticed the trails of molluscs, and the meandering furrows and ridges produced by worms, or annelides, and the tracks of crabs, and sometimes the footprints of birds and of dogs or other quadrupeds that have walked over sand or mud while it was yet plastic and sufficiently firm to retain the markings impressed upon it. Under certain conditions these apparently evanescent characters are indelibly fixed on the stratum, and in rocks of immense antiquity successive layers of sandstone and shale, through a thickness of many hundred feet, are found deeply furrowed with the ripples of the waves that flowed over them, pitted by the rain that has fallen upon them, and impressed with the footmarks of bipeds and quadrupeds that traversed the sands whilst the surface was in a moist and yielding state. Even on some of the most ancient of rocks, such as those of the Cambrian system, jelly-fish have left indelible impressions of their soft round bodies!

Speaking of the wonderfully enduring nature of certain impressions known to geologists, the sagacious Dean Buckland said, in an address to the Geological Society: "The historian or the antiquary may have traversed the fields of ancient or of modern battles, and may have pursued the line of march of triumphant conquerors, whose armies trampled down the most mighty kingdoms of the world. The winds and storms have utterly obliterated the ephemeral impressions in their course. Not a track remains of a single foot or a single hoof of all the countless millions of men and beasts whose progress spread desolation over the earth. But the reptiles that crawled upon the half-finished[1] surface of our infant planet have left memorials of their passage, enduring and indelible. Centuries and thousands of years have rolled away, between the time in which these footsteps were impressed by tortoises upon the sands of their native Scotland and the hour when they were again laid bare and exposed to our curious and admiring eyes. Yet we behold them stamped upon the rock, distinct as the track of the passing animal upon the recent snow; as if to show that thousands of years are but as nothing amidst eternity, and, as it were, in mockery of the fleeting perishable course of the mightiest potentates among mankind."[2]

Every form of animal life that, writhing, crawling, walking, running, hopping, or leaping, could leave a track, depression, or footprint behind it, might thereby leave similar lasting evidence of its existence and also, to some extent, of its nature. The

[1] This expression is a survival from the teaching in vogue fifty years ago. The world was not in an unfinished state during the period of the New Red Sandstone.

[2] *Bridgewater Treatise*, vol. i. p. 251.

interpretation of such evidences of ancient life has exercised the sagacity of naturalists since Dr. Duncan, in 1828, first inferred the existence of tortoises in certain sandstones in Dumfriesshire from the impressions left on them. The vast number and variety of such impressions has raised up a distinct branch of Palæontology, to which the name Ichnology [1] has been given.

We will now give a brief account of the results which have been arrived at in this branch of inquiry. To begin with one of the lowest forms of animal life—the worms. The class Annelida comprises the so-called ringed worms, including the leeches and earth-worms, and the sea-worms. As might have been expected, earth-worms are unknown in the geological record; for their soft bodies were not likely to be preserved even in the most favourable kinds of deposits. But in some cases the hard jaws of marine worms have escaped destruction. Fossil worm-jaws are abundantly found in some parts of the Cambrian, Silurian, and Carboniferous systems (see Appendix I.). The so-called "Conodonts" are believed by many authorities to be the jaws of worms; and such remains are also found in strata of the Mesozoic and Tertiary eras. Besides these rather mysterious little bodies, which some have taken to be the teeth of primitive fishes—such as our modern Hag-fish,—a good many worm-like markings are found in muddy and sandy sediments all through the stratified series. In many cases the true nature of these remains is still a matter of doubt. The visitor to the Natural History Museum at South Kensington will find in Gallery No. XI. a very fine and large collection of fossil tracks and footmarks of all kinds (Wall-cases Nos. 8–12). Some of these are probably vegetable remains; but others are certainly the tracks of molluscs or of crustaceans. Long burrows of marine worms occur plentifully in some rocks of Cambrian and Silurian

[1] Greek—*ichnos*, footstep; *logos*, discourse.

age, and have been figured under the names *Scolithus, Histioderma,* and *Arenicolites.* They are nearly straight, and descend vertically through the rock. Such are abundant in that ancient formation the Potsdam Sandstone of North America; in the Clinton formation, also of that country; and in the Stiperstones of Shropshire. Even in the Pre-Cambrian rocks of the west of Sutherlandshire there have been discovered of late years some long dark lines which are believed to be the burrows of marine worms pulled out to great lengths by the "shearing," or pulling-out process to which these rocks have been long ago subjected. For the full and complete interpretation of many of the curious markings known to geologists, a more accurate knowledge of the markings made by living animals will doubtless be necessary.

Various worms of the present day, such as the common lug-worm, are known to form long, wandering, tortuous channels in the sand of the seashore, a little distance below the surface. These worms feed on particles of organic matter scattered through the sand or mud, through which it eats its way. Such burrows cross and intersect each other in various ways, and as the worm proceeds on its course, they become filled up in the rear by the sand which has passed through its body. This is how worm-casts seen on the seashore at the present day are made. It appears, in the light of more recent researches, that many markings found in some of the more ancient (Palæozoic) rocks, and which have been formerly described as "fucoids," *i.e.* seaweeds, under such names as *Palæochorda,* are in reality the filled-up burrows of marine worms. These have now been re-christened *Planolites.* But there are some who consider them to be tracks made on the surface, *not* burrows.

A great many true trails, or tracks of worms, etc., that is markings made by the animal dragging its soft body over the surface of wet sand or mud, are found in the stratified rocks. But, in the present state of knowledge, it is very difficult to distinguish between those formed by worms and others made by molluscs or even crustaceans. The fossil known as Nereites, from Silurian slates at Wurtzbach, is probably the track of a worm. On the other hand, some of the tracks attributed to worms may have been really made by gastropod molluscs, such as whelks. One fossil track, known as Crossopodia, resembles the track made by a living Purpura lapillus, a well-known sea-shell. It is only in those strata which are very favourable to the preservation of organic remains that we can expect to find any trace or impression of the actual body of such a frail and perishable thing as a worm; but, incredible as it may seem, fossils of this nature occur in that most wonderful formation—the famous Solenhofen limestone, in which so many valuable treasures have been found; also in the Eocene slates of Monte Bolca (Italy). In these rare cases the form of the worm's body is actually seen, and the fossilised jaws occur in their natural position. Examples of these interesting specimens are beautifully rendered by chromolithography in the German monumental work *Palæontographica.*

Several geologists, such as Poulett-Scrope, Strickland, Buckland, Salter, and others, have published the results of careful comparisons of tracks made by living animals on the sands of the seashore, or on flat surfaces of mud left exposed by the drying up of a pond, or by other causes, but have not given drawings of the recent markings on which their conclusions are based. Professor Emmons and Professor T. McKenny Hughes, however, have figured some recent tracks in illustration of fossil ones. The former geologist came to the conclusion that certain imprints upon some very old rocks—the Taconic Shales of Maine and New

York States—were made by the soft fragile larvæ of insects which existed at an early period in the world's history.

We now pass on to the consideration of certain impressions in the old Potsdam Sandstone of North America, which have been most carefully studied by the late Sir R. Owen. In the year 1851, Logan exhibited before the Geological Society of London a small slab of sandstone showing some footprints, and a plaster cast from a longer surface of similar description. The original, weighing upwards of a ton, is in the Museum at Montreal connected with the McGill University The locality where it was found is on the left bank of the River St. Louis, at the village of Beauharnois, on the south side of the St. Lawrence, about twenty miles above the city of Montreal. Owen, in his first paper, came to the conclusion that the tracks were those of a tortoise. But further research caused him to alter this opinion. We only mention that to show how much care is required even on the part of the best naturalists to read the meanings of fossil tracks. A portion of the impressions now under consideration is shown in Fig. 1. They consist of a series of well-defined impressions continued in regular succession for four feet, and more; but only clearly for four feet. In this four feet there are thirty successive groups of footprints on each side of a furrow. The number of prints is not the same in each group. Where they are best marked, as in our figure, we see three prints in one group, two in the next and two in the third, followed by a repetition of the three prints (in our illustration each of the three groups is enclosed in an oval). These three groups (of 3-2-2 impressions) are distinctly repeated in succession along the whole series of tracks on both sides of the furrow. It will be noticed that in each pair of impressions the innermost pair are of equal size, but of the outer ones each is a little bigger than the last. An important point is that there are no marks of toes or nails. Their edges are not sharply defined, but are rounded off, and there is a slight variation in the form and depth of the corresponding impressions on each side of the furrow. But the reader will see from the figure that they do correspond with each other. Thus, take the three tracks at the bottom of the lower group on the right side of the furrow; the innermost of this group of three may

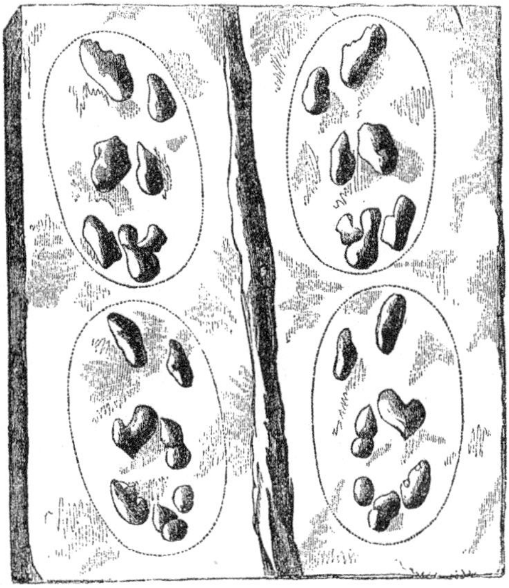

FIG. 1.—Tracks (*Protichnites*), probably of a crustacean, from the Potsdam Sandstone, North America. (After Owen.)

easily be identified with the innermost track of the group of three on the left side of the furrow. And so with the two groups of three belonging to the two upper sets of impressions, each of which is enclosed in an oval.

These very ancient tracks are known to geologists under the name Protichnites, and the creature that produced them must have made no less than fourteen impressions, seven on the right and seven on the left, each time its legs were set to work. There seems to be no doubt of this, because the groups of tracks, as marked out in our illustration, occur again and again in successive series so similarly and so regularly as to admit of no doubt that they were made by repeated applications of the legs, and these must have been capable of being moved so far in advance as to keep clear of the previous group of impressions. Sir Richard Owen concludes his account of the tracks by saying that the creature which made them was probably a crustacean genus, and that it may have had three pairs of limbs employed in locomotion, each of which was split up into two or more parts so as to make in walking either two or three tracks. The shape of the pits so clearly seen on these slabs of the old Potsdam Sandstone (although they have been rubbed and polished by the action of glacier-ice) suggests that they were made by the hard and partly pointed, partly blunt, terminations of the limb of a crustacean, such as a crab or lobster. But this creature moved directly forwards, not like a crab, but like a lobster or a king-crab. The furrow that runs between the tracks was probably made by a tail. The question then arises—what sort of a crustacean was it that made these tracks? Great caution is required in dealing with a problem of this kind, as will be seen from the following words of Sir R. Owen: "In all probability no living form of animal bears such a resemblance to that which the Potsdam footprints indicate as to afford an exact illustration of the shape and number of the instruments, and the mode of locomotion of the Protichnites." The imagination is baffled in the attempt to realise the extent of time past since the period when the creature was in existence which moved upon the sandy shores of the ancient Cambrian period, to which the "Potsdam Sandstone" belongs.

In about the year 1830, much interest was excited by the

discovery of footmarks, resembling those of land tortoises, on the exposed surfaces of slabs of sandstone of Triassic age, in a quarry at Corncockle Muir in Dumfriesshire, of which an interesting account was published by the Rev. Dr. Duncan. Regular tracks, indicating the slow progression of a small four-footed animal over the surface while the stone was in the state of moist sand, were traced on the blocks of sandstone, when separated by the quarrymen, along the lines of their stratification. In one instance there were found twenty-four consecutive impressions, forming a track with six distinct repetitions of the marks of each foot, the front feet differing from the hind feet. The appearance of five claws was discernible on the impressions of each fore paw. In 1853, Sir William Jardine published a splendid folio work in which he fully described these footprints; it was illustrated by full-sized lithographs coloured after Nature.[1]

The footprints occur in the Dumfriesshire Sandstones, in different patches, in several localities, but are best seen either where naturally exposed in the valleys of the Esk, the Nith, and the Annan, or in the quarries in those districts where they are worked for building material. One of those areas, of considerable extent, fills up the bottom of nearly all the upper basin of the Annan Valley above the ridge at Dormont Rocks. The beds are about two hundred feet thick, and present even surfaces. It is a curious fact, observed by the author of the above-mentioned book, that all the footprints are impressed as if the animal had walked from west to east. As a rule the creature seems to have walked in a straight line, but sometimes the tracks turn and wind in different directions. The paces are generally even and uninterrupted, seldom diverging much aside, showing little stoppage for food, or for a scuffle with a neighbour, which sometimes

[1] *Ichnology of Annandale* (Edinburgh, 1853; folio).

accompanied them. They appear more as the tracks of animals passing at once across some tide-receded estuary, in pursuit of some well-known and favourite grounds which were periodically sought after for some particular purpose. But it must be borne in mind that the impressions figured in this important work are not all similar in shape, and were probably due to different animals. They often show the effects of a peculiar pushing-back motion, which may be noticed in living tortoises. Dean Buckland, who was interested in these impressions, caused a living tortoise to walk on soft sand, clay, and paste, and found a fairly close correspondence between the tracks thus made and those of Corncockle Muir.

In 1831, Mr. Poulett-Scrope, an English geologist, described some small tracks made by a crustacean on a rock of the Jurassic period, known as the Forest Marble.

Some years previous to 1856, a series of strange impressions was found in a quarry in the lowest part of the Millstone Grit formation at Rhodes Wood, near Tintwhistle, Cheshire. The proprietor, Mr. Rhodes, was much struck with them, from the fact that they bore a resemblance to the marks of a human foot. The workmen also were struck with the resemblance, and, when they first showed him the impressions, remarked, "Master, some one has been here before us!" For several weeks the quarry was visited by many hundreds of people from Glossop and the surrounding neighbourhood. The common opinion was that the tracks were the footprints of some of Noah's family! This strange idea seems to have been founded on another equally strange, viz. that the Ark had rested on some neighbouring hills. But to return to the tracks; the distance between the impressions was two feet ten inches, and several of the impressions were thirteen inches long. Mr. Waterhouse Hawkins saw them, and thought that they resembled the supposed chelonian track figured by Jardine. Mr. Binney's conclusion (who read a paper on these tracks before the Geological Society) was that they were made on wet sand by a heavy slow-moving animal, like a tortoise, with irregular gait.

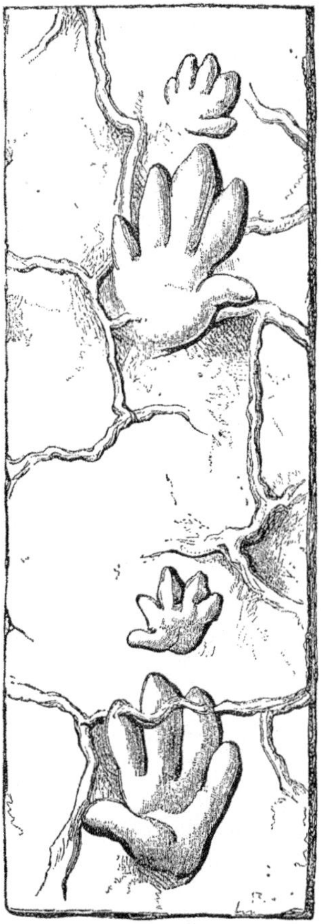

FIG. 2.—Footprints of *Cheirotherium*, in the Bunter Sandstone, Hessberg, near Hildburghausen.

Soon after the discovery of footprints at Corncockle Muir, another discovery was made in strata of the same geological age at Hessberg, near Hildburghausen, in Saxony. These footprints, however, were evidently made by somewhat large creatures, in which the fore paws were much smaller than the hind ones (see Fig. 2). Subsequently, similar tracks were observed on slabs of Triassic Sandstone in the quarries at Storton, in Cheshire. Others, again, have been found in Cheshire. The Museums at Warwick, Warrington, and Liverpool, are rich in impressed slabs from the New Red Sandstone. Numerous fine specimens may be seen in the Museum of the Geological Society, the Museum of Practical Geology, in Jermyn Street, and in the Natural History Museum, Cromwell Road, which is a part of the British Museum (Gallery No. XI.). Some of the slabs from Storton are covered with small round pits, or depressions, produced by rain-drops that fell while the surface was soft and impressible. The footprints from Storton are seen on the face of each successive stratum of marl, the corresponding surface of the overlying stone presenting, in relief, casts of the imprints and other markings. The hollow impressions of the feet are always on the upper surfaces of the slabs, and the convex casts on the under side of each layer or stratum, the latter fitting closely into the former. The double lines in between are casts of sun-cracks formed as the mud dried in the sun.

These "footprints on the sands of time" follow one another in pairs—one small, the other large, each pair being in the same line, and some fourteen inches in advance of the other. Each footmark has five toes, and the first, or little toe, is bent outward like a thumb, and is alternately on the right and left side of both the large and small footprints, which, except in size, resemble each other. The German geologist, Dr. Kaup, who first described them in 1834, proposed the name of *Cheirotherium*[1] for the great unknown animal that left the footprints, in consequence of the resemblance, both of the fore and hind feet, to the impression of a human hand. No *certain* remains of the creatures whose tracks we are now considering have yet been discovered in the same strata. But in these rocks and others of the same geological age in England and Germany there have been obtained skulls, teeth, and bones of amphibians, known as Labyrinthodonts, of which we shall have more to say in Chapter VI. Some of the salamander-like amphibians of the Triassic period, we now know from later discoveries, attained to a considerable size.

As Sir R. Owen points out, the impressions of the *Cheirotherium* resemble the footprints of a modern salamander in having the

[1] Greek—*cheir*, hand; *therion*, beast.

short outer toe of the hind foot projecting at right angles to the line of the middle toe, but yet are not identical with those of any known batrachian or reptile. Still it has been conjectured by the same great authority, as well as by others, that these footprints were the work of the creatures now known as Labyrinthodonts, which have left their remains in rocks of the Carboniferous, Triassic, and Permian ages. Later researches have shown that Owen was wrong. We need not be surprised at this, for Palæontology has, like all the other sciences, made great advances the last fifty years. The researches of Marsh, Cope, Huxley, Fritsch, Seeley, and others, have brought to light many new forms of old reptiles, some of which will be described in later chapters. The supposed Cheirotherium probably was one of the Dinosaurian reptiles, and in that case these footprints were made by Dinosaurs.

Our friend, Professor W. J. Sollas, has described some interesting footprints from South Wales, which probably were made by a Dinosaur of the Triassic age.[1] A friend of his was, in 1878, passing through the village of Newton Nottage, in Glamorganshire, when his attention was arrested by some three-toed footprints on a slab of rock, deeply impressed and rendered particularly visible by the slanting rays of the setting sun. Casts of them were afterwards made by the curator of the Cardiff Museum (Mr. J. Storrie).

To show how valuable geological finds are often neglected through ignorance of their real worth, it may be mentioned here that this slab even in 1894 was lying in a corner of the village green, in front of the church; formerly it lay in front of the steps of the inn, where it consequently suffered more or less wear. The impressions remind one of some of those described by Professor

[1] *Quarterly Journal of Geological Society*, xxxv. (1879), p. 510.

Hitchcock (see p. 45), and resemble more especially those belonging to his genus Brontozoum. Other specimens were afterwards found in the same locality. Professor Sollas had casts made of impressions of the feet or a living emu in the Clifton Zoological Gardens, for the sake of comparison, and found a good deal of agreement between the two. Nevertheless, from what we now know of Dinosaurs, it would be unwise to say that the impressions were made by birds.

Professor W. C. Williamson has described some very interesting impressions from Cheshire. They were found, by a former pupil of his (Mr. J. W. Kirkham), near Weston Point. They are unlike those of the Cheirotherium, previously described, and differ from all others yet found in showing very distinctly what are probably the *marks of scales.* The form of the foot also differs in being more quadrate. Professor Williamson says it reminds him of certain footprints found by Dr. King in the Carboniferous rocks of Pennsylvania. The arrangement of the scales corresponds closely with that seen on the foot of a modern alligator. The impressions suggest a saurian much more than an amphibian.

These impressions figured by Professor Williamson remind one a good deal of some tracks described by Professor Huxley, in his paper on a "New Red Sandstone Crocodile from the Traiassic Strata of Elgin,"[1] which may have been made by that ancient leviathan.

We now pass on to give some account of those famous footprints in the Connecticut Valley, of which probably all geologists have heard. The River Connecticut, in part of its course through the country which bears its name, and in the northern district of

[1] *Quarterly Journal of Geological Society*, xv. (1859), p. 440.

the adjoining State of Massachusetts, flows through a valley cut out of sandstone of the Triassic age. Successive layers of this rock are exposed all along considerable tracts of country. From this circumstance, and from the facility of transport afforded by the river, numerous quarries have, for many years, been worked in various parts of the valley, near the water's edge. The many footprints contained in these rocks were observed much earlier than the date (1828) in which the Rev. Dr. Duncan first described the tracks at Corncockle Muir (see p. 37). They have been very fully described and figured by Professor Ed. Hitchcock[1] and Dr. J. Deane.[2] As far as Dr. Deane could learn, the first specimen was ploughed up in South Harley, in 1802, by a boy. This specimen is now in the Appleton Ichnological Cabinet.[3] So strikingly did the tracks resemble those of birds, that they were familiarly spoken of as the tracks of poultry, or of "Noah's raven." It was not until the year 1836 that any attempt was made to describe the tracks scientifically. The year previous some flagging stones were obtained in Montague for the streets of Greenfield, by a Mr. Wilson, who observed impressions upon them, which he regarded as tracks of "the turkey tribe." These were observed by Dr. Deane, who sent casts of them to Professor Hitchcock. Professor Hitchcock gave his first account of them in *The American Journal of Science*, in 1836. He propounded the idea

[1] Professor Ed. Hitchcock, *Ichnology of New England.* (Boston, 1858.)

[2] Dr. J. Deane, *Ichnographs from the Sandstone of Connecticut River.* (Boston, 1861.)

[3] The late Honourable Samuel Appleton, of Boston, left by his will a large sum of money to be appropriated by the trustees under his will to benevolent and scientific purposes. Those trustees accordingly appropriated ten thousand dollars to the erection of the Appleton Cabinet at Amherst, a museum of which the lower story is entirely devoted to fossil footmarks (see *Report of Smithsonian Institution*, 1907). A still larger collection is now in the Museum of Yale University, but only a part is on view.

that they were the tracks of birds—a view which was not adopted by scientific men at the time, though afterwards many came round to his opinion. We have perused his work with great interest, and cannot but admire the care with which he studied

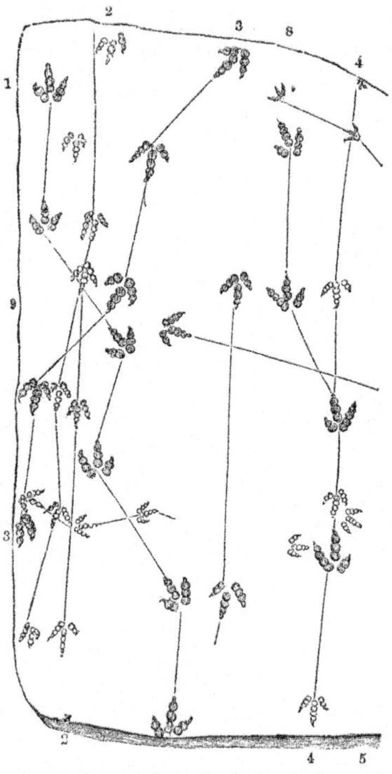

FIG. 3.—Portion of a slab of New Red Sandstone, from Turner's Falls, Massachusetts, U.S., covered with numerous tracks, probably of Dinosaurs. This specimen is now in the Natural History Museum, London. The separate tracks are indicated by the numbers. (After Hitchcock.)

the tracks and endeavoured to interpret their meaning, although his conclusions now require a good deal of modification.

Professor Hitchcock chose to give fanciful names to the creatures that made the tracks, such as Brontozoum giganteum,

1

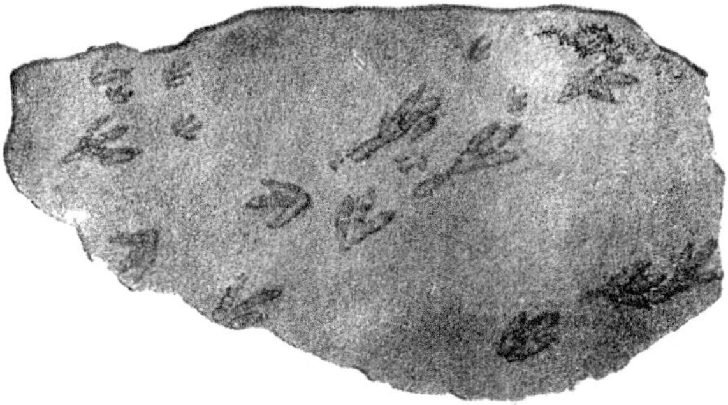

2

3

FOOTPRINTS OF DINOSAURS, FROM CONNECTICUT SANDSTONE.

PLATE I. Showing impressions of the small fore feet (1 and 3). (After Hitchcock.)

"the huge animal giant;" Polemarchus gigas, "the huge leader in war," and so on. He concludes a description of some of these creatures in the following words: "Such was the fauna of the sandstone days in the Connecticut Valley. What a wonderful menagerie! Who would believe that such a register lay buried in the strata? To open the leaves, to unroll the papyrus, has been an intensely interesting though difficult task, having all the excitement and marvellous developments of romance. And yet

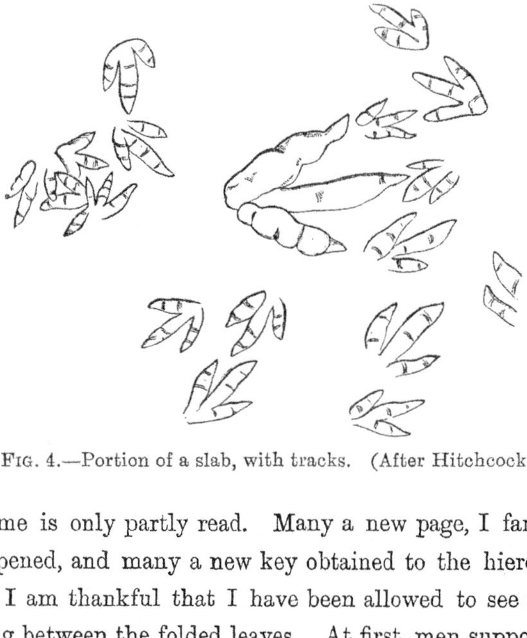

FIG. 4.—Portion of a slab, with tracks. (After Hitchcock.)

the volume is only partly read. Many a new page, I fancy, will yet be opened, and many a new key obtained to the hieroglyphic record. I am thankful that I have been allowed to see so much by prying between the folded leaves. At first, men supposed that the strange and gigantic races which I described were mere creatures of imagination, like the Gorgons and Chimæras of the ancient poets. But now that hundreds of their footprints, as fresh and distinct as of yesterday, impressed upon the mud, arrest the attention of the sceptic on the ample slabs of our cabinets, he might as reasonably doubt his own corporeal existence as that of these enormous and peculiar races."

Professor Hitchcock's work on this subject was much appreciated by his countrymen, as the following facts will show. It was resolved by the Commonwealth of Massachusetts, in the year 1857, that his *Geological Report on the Sandstone of the Connecticut Valley*[1] be printed, and one copy furnished to each member of the executive and legislative departments of the Government of that year, and one copy to each town and city in the Commonwealth. Not content with this, it was resolved in the following year that more copies be printed, that one hundred be given to the author, and others for the purpose of international exchange. Englishmen cannot but admire this good example, and would doubtless be glad if it could be followed at home, where so little is done for the cause of Science by our statesmen.

But, in justice to Dr. Deane, we must not omit to mention his part in the description and figuring of these interesting records. Unfortunately, he died before the completion of his beautifully illustrated work on the subject. He drew with his own hand a large number of lithographs from the actual slabs; and these drawings will always be of the greatest use to students of Ichnology. He worked at the subject in his spare moments, during an active professional career, and with great enthusiasm, as the following extract from his unfinished work will show.

"An indescribable interest is imparted by opening the long-sealed volume that contains the records of these extinct animals. The slabs were uncovered and raised under my supervision, and page after page, with their inscriptions, revealed living truths. There were the characters, fresh as upon the morning when they were impressed, reminding the spectator of the brevity of human

[1] *Ichnology of New England.*

antiquity, and of the frail tenure of human works! On that morning, how long ago no one can tell, or ever will know, gentle showers watered the earth, an ocean was unruffled, and upon its borders primæval beings enjoyed their existence and inscribed their eventful history."

And now to sum up the results of their work. Most of the tracks so fully described and figured by these two authors were probably made by amphibians and reptiles, and it is doubtful if any of them were made, as they supposed, by birds. They vary extremely, both in size and character. While some are only half an inch long; others, like the huge Otozoum, of Hitchcock (see Fig. 5, left-hand corner), are twenty inches long, and show a stride of three feet! Some of the creatures that made the tracks had five toes, some four, and many of them only three. Again, some had hind feet and fore feet of nearly equal size, and evidently walked or crawled on all-fours.

We have here reproduced, from Professor Hitchcock's work, a highly interesting plate (Plate I.), which shows that some of the animals had large hind feet and small fore feet, and that they sometimes put down the latter, as kangaroos do, for there are the impressions of them. However, such impressions are rare, and, as a rule, the creatures, many of which were probably Dinosaurs, walked on their hind limbs, thus producing those three-toed bird-like impressions which are shown on the slabs represented in our figures. It is hardly likely that they hopped on their hind legs as kangaroos do; they were probably too heavy for that; but they may have made something like a hop in bringing up their hind feet, as we have seen the kangaroos do in the Zoological Gardens. Another very interesting point brought to light by the researches of Hitchcock and Deane is, that on some slabs may be seen what is believed to be the impression of the tail of a Dinosaur as it dragged along over the moist sand of the seashore or estuary where these antediluvian creatures of old disported themselves (see Fig. 5).

Those impressions which Hitchcock believed to have been made by birds show a regular increase in the number of joints

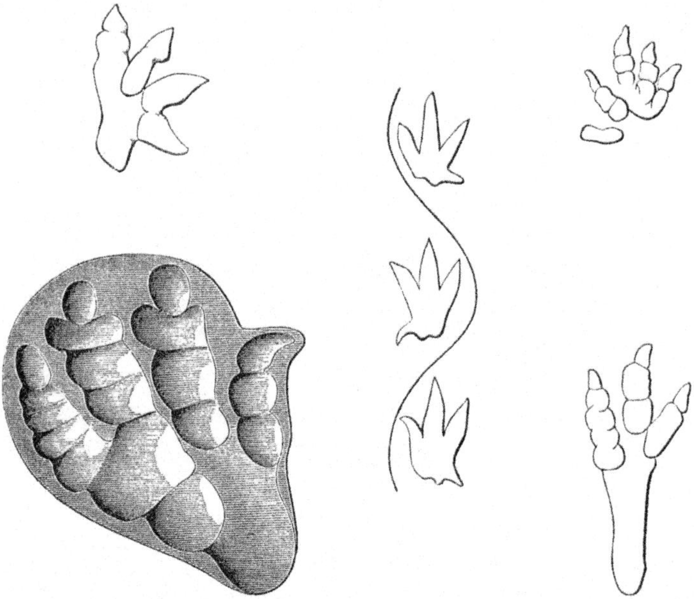

FIG. 5.—Footprints from the Connecticut Sandstone. The three tracks in the middle show the mark left by the tail, probably of a Dinosaur (After Hitchcock.)

of the toes; the inner toe having two, the middle one three, and the outer one four joints. Now, it happens, by a remarkable coincidence, that in the case of birds the inner toes have three, the middle toes four, and the outer toes five joints, but the last two joints in each case make but one division of the track, so that their tracks correspond with those we are considering. The discovery that some Dinosaurs have but three toes that were used (functional toes), and that they had the same number of joints in their toes as birds, has naturally led scientific men to the opinion that probably *all* the tracks above described were made, not by birds, but by Dinosaurs. It may be pointed out in defence of this opinion, that, although the two classes of Birds and Reptiles are *now* widely separated, yet in certain former geological periods there was no such gulf as now divides them. At the time when the Connecticut Sandstone tracks were made on the shores of a narrow inland sea, there must have been in existence animals which, if we saw them now, might sorely puzzle us to decide whether to call them reptilian birds, or bird-like reptiles.

During the Jura-Trias period there was in the region of the Connecticut Valley a shallow sea, connected by a narrow outlet with the ocean. Into this the tides flowed and ebbed, leaving extensive flats of mud or sand, ribbed with ripple-marks. A passing shower pitted the soft mud, and the sun, coming out again from the breaking clouds, dried and cracked it. Our Dinosaurs and other creatures sauntered or ran near the margin of the shore. The tide came in again, carrying with it fine sediments, gently covered the tracks, and preserved them for ever. This occurred constantly for many ages, about the time when the Triassic period came to a close.

In the year 1882, reports were published of the discovery of large footprints—supposed to be human—in a certain sandstone, near Carson, Nevada, U.S., of which a brief account was given in our former work.[1] These are probably the tracks made by a big extinct sloth of the Pleistocene period. The wonderful series of footprints of reptiles, birds, and mammals discovered by M. Desnoyers, in certain Eocene strata near Paris, cannot be described here.

[1] *Extinct Monsters*, p. 185 (2nd edit.).

CHAPTER III

SEA-SCORPIONS

"And some rin up the hill and down dale, knapping the chucky stanes to pieces wi' hammers like sae many road-makers rin daft. They say 'tis to see how the warld was made."—Scott, *St. Ronan's Well.*

OUR first group of monsters is taken from a tribe of armed warriors that lived in the seas of a very ancient period in the world's history. Like the crabs and lobsters inhabiting the coasts of Britain, they possessed a coat of armour, and jointed bodies, supplied with limbs for crawling, swimming, or seizing their prey. They were giants in their day, far eclipsing in size any of their relations that have lived on to the present time. Some of them, such as the Pterygotus (Fig. 6, p. 51), attained a length of nearly six feet.

There are no living creatures quite like them. Certainly they are not true lobsters, and yet we may consider them to be first or second cousins of those ten-footed crustaceans[1] of the present day—lobsters, crabs, and shrimps, so welcome on the tables of both rich and poor. Some naturalists say that their nearest relations at the present day are the king-crabs inhabiting the China seas and the east coast of North America; and there certainly are some points of resemblance between them. Others say

[1] Crustaceans are a class of jointed creatures (articulate animals), possessing a hard shell or crust (Lat. *crusta*), which they cast periodically. They all breathe by gills.

that they are related to scorpions, and for this reason we call them Sea-scorpions. (See Plate II.)

The Crustacea are a very ancient race, as well as a flourishing one; some very old rocks containing undoubted fossils—namely, certain slates found in Wales and the Lake District—tell us of a

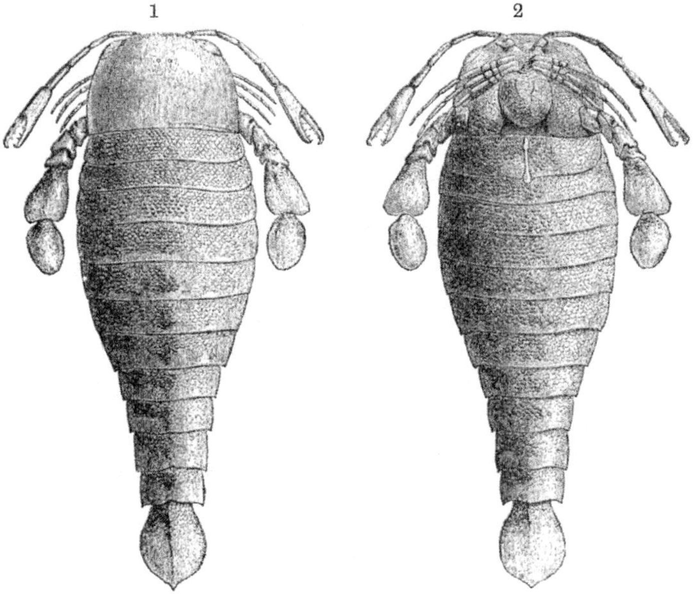

FIG. 6.—*Pterygotus anglicus*, six feet long.
1. Upper side. 2. Under side. (After H. Woodward.)

time when shallow seas swarmed with little articulate animals known as *trilobites.*

The best-known and largest of the Sea-scorpions is represented in Fig. 6. It has received the name Pterygotus (or wing-eared), from certain fanciful resemblances pointed out by the quarrymen. It was first discovered, along with others of its kind, by Hugh Miller, at Carmylie in Forfarshire, in a certain part of the Old Red Sandstone (see Table of Strata, Appendix I.) known as the

Arbroath paving-stone. The quarrymen, in the course of their work, came upon and dug out large pieces of the fossilised remains of this creature. Its hard coat of jointed armour bore on its surface curious wavy markings that suggested to their minds the sculptured feathers on the wings of cherubs—of all subjects of the chisel the most common. Hence they christened these remains "Seraphim." They did not succeed in getting complete specimens that could be pieced together; and the part to which this fanciful name was given turned out to be part of the under side below the mouth. It was composed of several large plates, two of which are not unlike the wings of a cherub in shape. Hugh Miller says in his classic work, *The Old Red Sandstone*—"the form altogether, from its wing-like appearance, its feathery markings, and its angular points, will suggest to the reader the origin of the name given it by Forfarshire workmen."

A correct restoration, in proportion to the fragments found in the Lower Old Red Sandstone, would give a creature measuring nearly six feet in length, and more than a foot across. Pterygotus anglicus may therefore be justly considered a monster crustacean.

The illustrious Cuvier, who, in the eighteenth century, founded the science of comparative anatomy (see p. 5), astonished the scientific world by his bold interpretations of fossil bones. From a few of the chief bones he could restore the skeleton of an entire animal, and determine its habits and mode of living. When other wise men were unable to read the writing of Nature on the walls of her museum—in the shape of fossil bones—he came forward, like a second Daniel, to interpret the signs, and so instructed us how to restore the world's lost creations. Hugh Miller submitted the fragments found at Balruddery to the celebrated naturalist Agassiz, a pupil of Cuvier, who had written a

SEA-SCORPIONS.

Pterygotus anglicus. Length 6 feet.

Eurypterus.

Stylonurus.

PLATE II.

famous work on fossil fishes; and he says that he was much struck with the skill displayed by him in piecing together the fragments of the huge Pterygotus. "Agassiz glanced over the collection. One specimen especially caught his attention—an elegantly symmetrical one. His eye brightened as he contemplated it. 'I will tell you,' he said, turning to the company—'I will tell you what these are—the remains of a huge lobster.' He arranged the specimens in the group before him with as much ease as I have seen a young girl arranging the pieces of ivory in an Indian puzzle. There is a homage due to supereminent genius, which Nature spontaneously pays when there are no low feelings of jealousy or envy to interfere with her operations; and the reader may well believe that it was willingly rendered on this occasion to the genius of Agassiz." Agassiz himself, previous to this, had considered such fragments as he had seen to be the remains of fishes. As we have said before, this creature was *not* a true lobster; but Agassiz, when he expressed the opinion just quoted, was not far off the mark, and did great service in showing it to be a crustacean. There were no lobsters at that early period of the world's history, and this creature, with its long "jaw-feet" and powerful tail, was a near approach to a king-crab on the one hand and scorpion on the other. If living now, it would no doubt command a high price at Billingsgate; but, then, it would be a dangerous thing to handle when alive, and might be more troublesome to catch than our crabs or lobsters.

The front part of the body was entirely enveloped in a kind of shield, called a carapace, bearing near the centre minute eyes, which probably were useless, and at the corners two large compound eyes, made up of numerous little lenses, such as we see in the eye of a dragon-fly. This is clearly proved by certain well-preserved specimens. There are five pairs of appendages, all attached under or near the head. Behind the head follow twelve rings, or segments, the last of which forms the tail, two at least of these bore gills for breathing. All but two of them, below the mouth, must have been beautifully articulated, so as to allow them to move freely, as we see in the lobster of the present day. But look at that lowest and largest pair of appendages, the end joints of which are flattened out, and you will see that they must have been a powerful oar-like apparatus for swimming forwards. We can fancy this creature propelling itself much in the same way as a "water-beetle" rows itself through the water in a pond. In all other crustaceans the antennæ are used for feeling about, but in the Pterygotus they are used as claws for seizing the prey.

From its large eyes, powerful oar-like limbs, or appendages, and from the general form of its body, Dr. Henry Woodward (the author of a learned monograph on these creatures) concludes that the Pterygotus was a very active animal; and the reader will easily gather from its pair of antennæ, converted at their extremities into nippers, and from the nature of its "jaw-feet," that the creature was a hungry and predaceous monster, seizing everything eatable that came in its way. The whole family to which it belongs—including Pterygotus, Eurypterus, Slimonia, Stylonurus, and others—seems to have been fitted for rather rapid motion, if we may judge from the long tapering and well-articulated body. In two forms (Pterygotus and Slimonia) the tail-flap probably served both as a powerful propeller, and as a rudder for directing the creature's course; but others, such as Eurypterus and Stylonurus, had long sword-like tails, which may have assisted them to burrow into the sand, in the same way that king-crabs do. Eurypterus remipes is shown in Fig. 7. A beautiful restoration in wax of Eurypterus Fischeri may be seen in the Gallery of Crustacea in the Natural History Museum.

Our Sea-scorpions are related to the king-crab of the present day. Now, this creature, it is well known, burrows into the mud and sand at the bottom of the sea. This it does by shoving its broad sharp-edged head-shield downwards, working rapidly at the same time with its hinder feet, or appendages, and by pushing with the long spike that forms a kind of tail. It will thus sink deeper and deeper until nothing can be seen of its body, and only the eyes peep out of the mud. It will crawl and wander about by night, but remains hidden by day. Some individuals are of large size, and occasionally measure two feet in length. They possess six pairs of well-formed feet, the joints of which, near the body, are armed with teeth and spines, and serve the purpose of jaws, being used to masticate the food and force it into the mouth, which is situated between them.

Now, this fact is of great importance; for it helps us to understand the use of the four pairs of "jaw-feet" in our Sea-scorpions. What curious animals they must have been, using the same limbs for walking, holding their prey, and eating! Look at the broad plates at the base of the oar-like limbs, or appendages, with their tooth-like edges. These are the plates found by Hugh Miller's quarrymen, and compared by them to the wings of seraphim. The reader will easily perceive that by a backward and forward movement, they would perform the office of teeth and jaws, while the long antennæ with their nippers—helped by the other and smaller appendages—held the unfortunate victim in a relentless grasp. And even these smaller limbs, as seen from the figure, had their first joints, near the mouth, provided with toothed edges like a saw.

With regard to the habits of Sea-scorpions, it would not be

altogether safe to conclude that, because in so many ways they resembled king-crabs, they therefore had the same habit of burrowing into the soft muddy or sandy bed of the sea, as some authorities have supposed. Seeing that there is a difference of

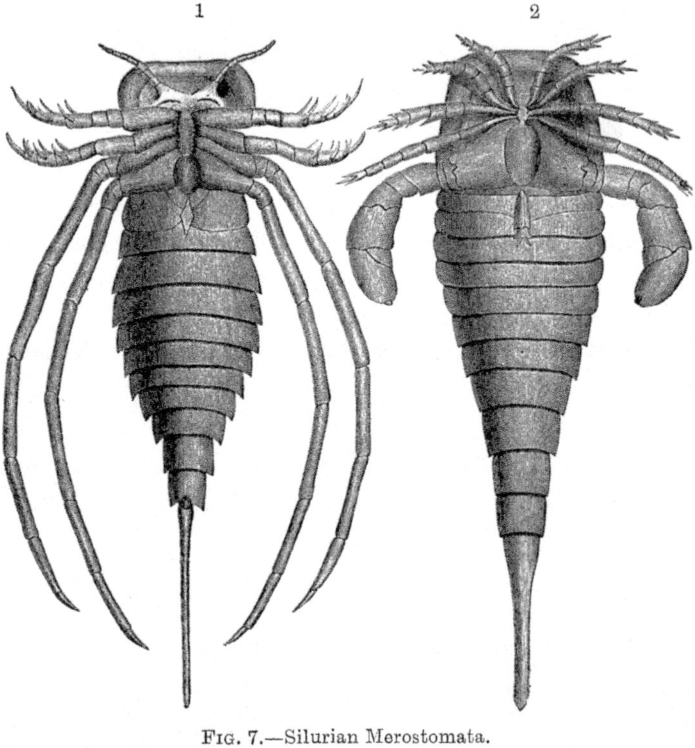

Fig. 7.—Silurian Merostomata.
1. *Stylonurus*. 2. *Eurypterus*. (After H. Woodward.)

opinion on this subject, the author consulted Dr. H. Woodward on the question, who said he thought it unlikely, seeing that, in some of them, such as the Pterygotus, the eyes are placed on the margin of the head-shield; for it would hardly care to rub its eyes with sand. Whether it chose at times to bury its long body in the sand by a process of wriggling backwards, as certain modern crustaceans do, we may consider to be an open question.

If only Sea-scorpions had not unfortunately died out, how interesting it would be to watch them alive, and to see exactly what use they would make of their long bodies, tail-flaps, and tail-spikes! Were they nocturnal in their habits, wandering about by night, and taking their rest by day? Such questions, we fear, can never be answered. But their large eyes would have been able to collect a great deal of light when the moon and stars feebly illumined the shallower waters of the seas of Old Red Sandstone times; and so there is nothing to contradict the idea.

Now, one of the many points of interest in Pterygotus and its allies is that they somewhat resemble the crab in its young or larval state. To a modern naturalist, this fact is important as showing that crustacean forms of life have advanced since the days of the sea-scorpions.

Their resemblance to land-scorpions is so close that, if it were not for the important fact that scorpions breathe *air* instead of *water*, and for this purpose are provided with an air-tube (or trachea) such as all insects have, they would certainly be removed bodily out of the crustacean class, and put into that in which scorpions and spiders are placed, viz. the Arachnida. But, in spite of this important difference, there are some naturalists in favour of such a change. It will thus be seen that our name Sea-scorpions is quite permissible. Insects and Crustacea are now included in the term Arthropoda.

Hugh Miller described some curious little round bodies found with the remains of the Pterygotus, which it was thought were the eggs of these creatures!

Finally, these extinct crustaceans flourished in those ages of the world's history known as the Silurian and the Old Red Sandstone periods. As far as we know, they did not survive beyond the succeeding period, known as the Carboniferous.[1]

[1] The student should consult Dr. Henry Woodward's valuable *Monograph of the British Merostomata* (Palæontographical Society), to which the writer is much indebted. With regard to the representation of *Pterygotus anglicus* in Plate II., it has been pointed out by Dr. Woodward that the creature was unable to bend its body into such a position as is shown there. As in a modern lobster or shrimp, there were certain overlapping plates in the rings, or segments, of the body, which prevented movement from side to side, and only allowed of a vertical movement.

CHAPTER IV

THE GREAT FISH-LIZARDS

"Berossus, the Chaldæan saith: A time was when the universe was darkness and water, wherein certain animals of frightful and compound forms were generated. There were serpents and other creatures with the mixed shapes of one another, of which pictures are kept in the temple of Belus at Babylon."—*The Archaic Genesis.*

VISITORS to Sydenham, who have wandered about the spacious gardens so skilfully laid out by the late Sir Joseph Paxton, will be familiar with the great models of extinct animals on the "geological island." These were designed and executed by that clever artist, Mr. Waterhouse Hawkins, who made praiseworthy efforts to picture to our eyes some of the world's lost creations, as restored by the genius of Sir Richard Owen and other famous naturalists. His drawings of extinct animals may yet be seen hanging on the walls of some of our provincial museums; but of these many are far from correct. The difficulties were much greater in those early days. Mr. Karl Hagenbeck is now setting up "Restorations" in his famous Zoological Garden at Hamburg.

Lazily basking in the sun, when it condescends to shine, and resting his clumsy carcase on the ground that forms the shore near the said geological island at Sydenham, may be seen the old fish-lizard, or Ichthyosaurus, that forms the subject of the present chapter. He looks awkward on land, as if longing to get into his native element once more, and cleave its waters with his powerful tail-fin. His "flippers" seem too weak to enable him to crawl on land. Moreover, the most recent discoveries of Dr. Fraas lead us to conclude that the Ichthyosaur never ventured to leave the "briny ocean" to bask upon the land.

This great uncouth beast presents some curious anomalies in his constitution, being planned on different lines to anything now living, and presenting, as so many other extinct animals do, a mixture, or fusion, of types that greatly puzzled the learned men of the time when his remains were first brought to light, after their long entombment in the Lias rocks forming the cliffs on the coast of Dorset. Some have christened him a "sea-dragon," and such indeed he may be considered. But the name Ichthyosaurus, given above, has received the sanction of high authority, and, moreover, serves to remind us of the fact that, although in many respects a lizard, he yet retains in his bony framework the traces of a remote fishy ancestry. So we will call him a fish-lizard.

The curious quotation given at the head of the present chapter refers to a widespread belief, prevalent among the highly civilised nations of antiquity, that the world was once inhabited by dragons, or other monsters "of mixed shape" and characters. To the student of ancient history traces of this curious belief will be familiar. Sir Charles Lyell refers to such a belief when he says, in his *Principles of Geology*, "The Egyptians, it is true, had taught, and the Stoics had repeated, that the earth had once given birth to some monstrous animals that existed no longer." It may be surprising to some, but it is undoubtedly the fact, that modern scientific truths were partly anticipated by the civilised nations of long ago.

The illustrious Cuvier, in his day, considered the fish-lizard to be one of the most heteroclite and monstrous animals ever discovered (see p. 74). He said of this creature that it possessed the snout of a dolphin, the teeth of a crocodile, the head and

SKELETON OF ICHTHYOSAURUS TENUIROSTRIS IN THE NATURAL HISTORY MUSEUM, LONDON.

PLATE III.

breast-bone of a lizard, the paddles of a whale or dolphin, and the vertebræ of a fish! No wonder that naturalists and palæontologists, whose realm is the natural history of the past, were obliged to make a new division, or order, of reptiles to accommodate the fish-lizard. It is obvious that a creature with such very "mixed" relationships would be out of place in any of the four orders into which living reptiles, as represented by turtles, snakes, lizards, and crocodiles, are divided. Here is what our late friend Professor Blackie said of the Ichthyosaurus—

"Behold, a strange monster our wonder engages!
If dolphin or lizard your wit may defy.
Some thirty feet long, on the shore of Lyme-Regis,
With a saw for a jaw, and a big staring eye.
A fish or a lizard? An ichthyosaurus,
With a big goggle eye, and a very small brain,
And paddles like mill-wheels in chattering chorus,
Smiting tremendous the dread-sounding main."

A glance at our restoration, Plate IV., will show that the fish-lizard was a powerful monster, well endowed with the means of propelling itself rapidly through the water as it sought its living prey, to seize it within those cruel jaws. The long and powerful tail was its chief organ of propulsion; but the paddles would also be useful for this purpose, as well as for guiding its course. The pointed head and generally tapering body suggests a capability of rapid movement through the water; and since we know for certain that it fed on fishes, this conclusion is confirmed, for fishes are not easily caught now, and most probably were not easily caught ages ago.

The personal history of the fish-lizard, merely as a fossil or "remain" is interesting; so much so, that we may perhaps be allowed to relate the circumstances of his *début* before the scientific world, in the days of the ever-illustrious Cuvier, to whom we have already alluded. But England had its share of illustrious men, too, though lesser lights compared to the founder of comparative anatomy,—such as Sir Richard Owen, on whom the mantle of his friend Cuvier has fallen; Conybeare, De la Beche, Dean Buckland, and Huxley.

These scientific men, aided by the untiring labours of many enthusiastic collectors of organic remains, have been the means of solving the riddle of the fish-lizard, and of introducing him to the public. By this time there is, perhaps, no creature among the host of Antediluvian types better known than this reptile.

The remains of fish-lizards have attracted the attention of collectors and describers of fossils for nearly two centuries past. The vertebræ, or "cup-bones," as they are often called, of which the spinal column was composed, were figured by Scheuchzer, in an old work entitled *Querelæ Piscium;* and, at that time, they were supposed to be the vertebræ of fishes. In the year 1814 Sir Everard Home described the fossil remains of this creature, in a paper read before the Royal Society, and published in their *Philosophical Transactions.* This fossil was first discovered in the Lias strata of the Dorsetshire coast. Other papers followed till the year 1820. We are chiefly indebted to De la Beche and Conybeare for pointing out and illustrating the nature of the fish-lizard; and that at a time when the materials for so doing were far more scanty than they are now. Mr. Charles König, Mr. Thomas Hawkins, Dean Buckland, Sir Philip Egerton, and Professor Owen have all helped to throw light on the structure and habits of these old tyrants of the seas of that age, which is known as the Jurassic period. They lived on, however, to the succeeding or Cretaceous period, during which our English chalk was forming; but the Liassic age was the one in which they flourished most abundantly, and developed the greatest variety.

In the year 1814 a few bones were found on the Dorsetshire coast between Charmouth and Lyme Regis, and added to the collection of Bullock. They came from the Lias cliffs, undermined by the encroaching sea. Sir Everard's attention being attracted to them, he published the notices already referred to. The analogy of some of the bones to those of a crocodile, induced Mr. König, of the British Museum, to believe the animal to have been a saurian, or lizard; but the vertebræ, and also the position of certain openings in the skull, indicated some remote affinity with fishes, but this must not be pressed too far. The choice of a name, therefore, involved much difficulty; and at length he decided to call it the Ichthyosaurus, or fish-lizard. Mr. Johnson, of Bristol, who had collected for many years in that neighbourhood, found out some valuable particulars about these remains. The conclusions of Dean Buckland, then Professor of Geology at Oxford, led Sir Everard to abandon many of his former opinions. The labours of the learned men of the day were greatly assisted by the exertions of Miss Anning, an enthusiastic collector of fossils. This lady, devoting herself to science, explored the frowning and precipitous cliffs in the neighbourhood of Lyme Regis, when the furious spring-tide combined with the tempest to overthrow them, and rescued from destruction by the sea, sometimes at the peril of her life, the few specimens which originated all the facts and speculations of those persons whose names will ever be remembered with gratitude by geologists.

Probably our readers are already more or less familiar with the drawings of the fossilised remains of Ichthyosauri to be seen in almost every text-book of geology. (Fig. 8 is from Owen's *British Fossil Reptiles.*) But we recommend all who take an interest in the world's lost creations to pay a visit to the great Natural History Museum, at South Kensington. The fossil reptile gallery contains a magnificent series of Ichthyosauri, about thirty in number. Of these a large number were obtained through the exertions of the late Mr. T. Hawkins, a Somersetshire gentleman, who was a most ardent collector of fossil reptiles, and who devoted himself with great enthusiasm and unsparing energy to the acquisition of a truly splendid collection of these most interesting relics of the past. Nearly seventy years ago he arranged for the purchase of his treasures by the authorities of the British Museum, and thus his collection became the property of the nation.

His specimens were figured and described by him in two large folio volumes. The first was published in 1834, under the title,

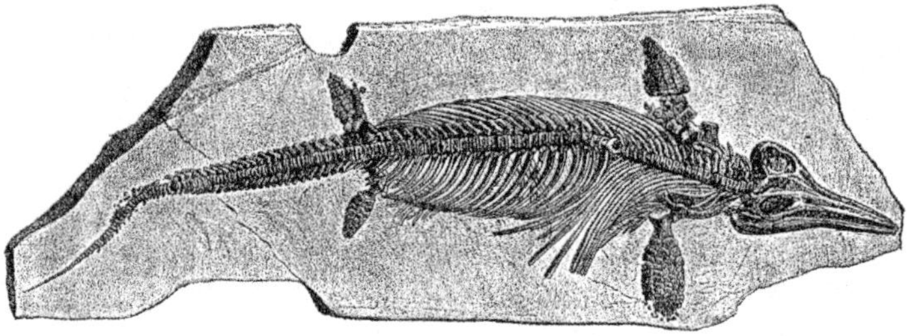

FIG. 8.—*Ichthyosaurus intermedius.*

Memoirs of the Ichthyosauri and Plesiosauri; his second, with the same plates, in 1842, under the quaint title of *The Book of the Great Sea-Dragons.* The large lithographic drawings of his fine specimens were beautifully executed by Scharf and O'Neil. The plates are the only really valuable part of these two curious and ill-written books.

Mr. Hawkins tells us that his youthful attention was directed to the Lias quarries, near Edgarly, in Somersetshire, in consequence of some strange reports. It was said that the bones of giants and infants had, at distant intervals, been found in them. These quarries he visited, and, by offers of generous payment, induced

PLATE IV. *Ichthyosaurus communis.* FISH LIZARDS. LIAS PERIOD. FISHES (*Dapedius, etc.*). *Ichthyosaurus tenuirostris* (a smaller species).

the workmen to keep for him all the remains they might find. In this way he finally obtained the co-operation of all the quarrymen in the county.

When, however, he complains of the Philistine dulness and stupidity of quarrymen, who often, in their ignorance, break up finds of almost priceless value, we can fully sympathise.

In general contour the body of the fish-lizard was long and tapering, like that of a whale (see Plate IV.). It probably showed no distinct neck. The long tail was its chief organ of propulsion. We notice two pairs of fins, or paddles; one on the fore part of the body, the other on the hinder part, like the pectoral and abdominal fins of a fish. The skin was scaleless and smooth, or slightly wrinkled, like that of a whale. No traces of scales have ever been found; and if such had existed, they would certainly have been preserved, since those of fishes and crocodiles of the Jurassic period have been found in considerable number and variety. It is therefore safe to conclude that such were absent in this case. In the Lias strata, at least, the specimens are often preserved with most wonderful completeness (see p. 70).

The long and pointed jaws are a striking feature of these animals. The eyes were very large and powerful, and specially adapted, as we shall see presently, to the conditions of their life.

It might, perhaps, be asked whether the fish-lizards breathed, like fishes, by means of gills. That question can easily be answered; for if they had possessed gills for taking in water and breathing the air dissolved therein, they would reveal the fact by showing a bony framework for the support of gills, such as are to be found in all fishes. These structures, known as "branchial arches," are absent; therefore the fish-lizards possessed lungs, and breathed air like reptiles of the present day. Their skulls show where the nostrils were situated; namely, near the eyes, and not at the end of the upper jaw-bone. There are also passages in the skull leading from the nostrils to the palate, along which currents of air passed on their way to the lungs. Being air-breathers, they would be compelled occasionally to seek the surface of the sea, in order to obtain a fresh supply of the life-giving element—oxygen; but, being cold-blooded and with a small brain, needing a much less supply of oxygen for its work, the fish-lizards had, like fishes,

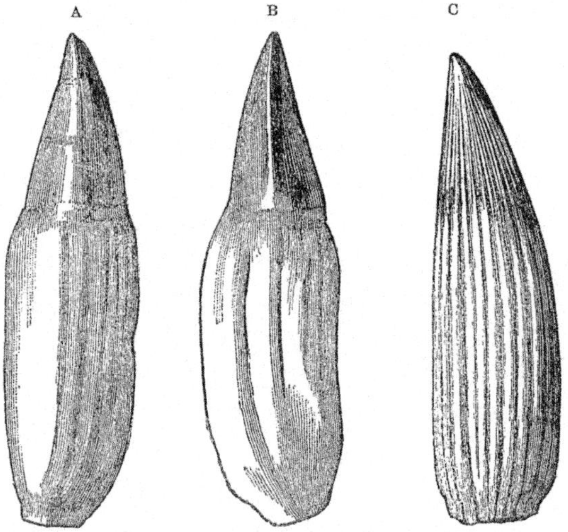

FIG. 9.—(A) Lateral and (B) profile views of a tooth of *Ichthyosaurus platyodon* (Conybeare), Lower Lias, Lyme Regis, Dorsetshire. (C) Tooth of *Ichthyosaurus communis* (Conybeare), Lower Lias, Lyme Regis, Dorset.

this advantage over whales, which are warm-blooded—that their stern-propeller, or tail-fin, could take the form best adapted for a swift, straightforward course through the water.

In the whale tribe the tail-fin is horizontal; and this is so on account of their need, as large-brained, warm-blooded air-breathers, of speedy access to the atmospheric air. Were it otherwise, they would not have the means of rising with sufficient rapidity to the surface of the sea; for they have only one pair of fins. But the fish-lizards had two pairs of these appendages, and the hinder or pelvic pair no doubt were of great service in helping the creatures to come up to the surface when necessary.

These great marine predaceous reptiles literally swarmed in the seas of the Lias period, and no doubt devoured immense shoals of the fishes of those times, whose numbers were thus to some extent kept down. There is the clear proof of this in the fossilised droppings—known as "coprolites,"—which show on examination the broken and comminuted remains of the little bony plates of ganoid fishes that we know were contemporaries of these reptiles. Probably young ones were sometimes devoured too (see p. 72).

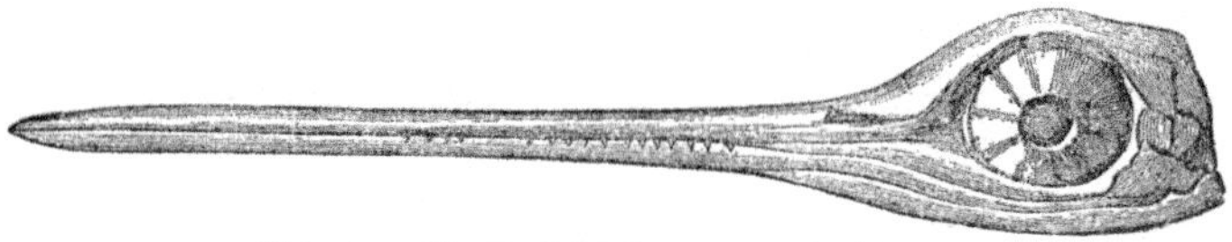

FIG. 10.—Skull of *Ichthyosaurus latifrons*.

It was in the period of the Lias that fish-lizards attained to their greatest development, both in numbers and variety; and the strata of that period have preserved some interesting variations. It will be sufficient here to point out two, namely, Ichthyosaurus tenuirostris—an elegant little form, in which the jaws, instead of being massive and strong, were long and slender like a bird's beak; and also Ichthyosaurus latifrons (Fig. 10), with jaws still more birdlike. Our artist has attempted to show the former variety in our illustration (Plate IV.). A most perfect example of this pretty little Ichthyosaur, from the Lower Lias of Street in Somerset, has been presented to the National Collection at South Kensington by Mr. Alfred Gillett, of Street, and may be seen there (Plate III.). In this group of fish-lizards the eyes are relatively larger, and we should imagine that they were very quick in detecting and catching their prey; their paddles also have larger bones.

There is a remarkably fine specimen at Burlington House, in the rooms of the Geological Society, of an Ichthyosaurus' head, which the writer found, on measuring, to be about five feet six inches long. A cast of this head is exhibited at the Natural History Museum. The largest of the specimens in the National Collection is twenty-two feet long and eight feet across the expanded paddles; but it is known that many attained much greater dimensions. Judging from detached heads and parts of skeletons, it is probable that some of them were between thirty and forty feet long. A specimen of Ichthyosaurus platyodon in the collection of the late Mr. Johnson, of Bristol, has an eye-cavity with a diameter of fourteen inches. This collection is now dispersed.

With regard to their habits, Sir Richard Owen concludes that they occasionally sought the shores, crawled on the strand, and basked in the sunshine. His reason for this conjecture (which, however, is not confirmed by Dr. Fraas's recent discoveries) is to be found in the bony structure connected with the fore paddles, which is not to be found in any porpoise, dolphin, grampus, or whale, and for want of which these creatures are so helpless when left high and dry on the shore. The structure in question is a strong bony arch, inverted and spanning across beneath the chest from one shoulder to the other. A fish-lizard, when so visiting the shore for sleep, or in the breeding season, would lie or crawl,

[1] It is, perhaps, hardly necessary to remark that whales are not fishes but mammals which have undergone great change in order to adapt themselves to a marine life. Their hind limbs have practically vanished, only a rudiment of them being left. But their external resemblance to the Ichthyosaur type is remarkable—dolphins especially so (see p. 73).

prostrate, with its under side resting or dragging on the ground—somewhat after the manner of a turtle.

It is a curious fact that this bony arch resembles the same part in those singular and problematical mammals, the Echidna and the Platypus, or duck-mole.

The enormous magnitude and peculiar construction of the eye are highly interesting features. The expanded pupil must have allowed of the admittance of a large quantity of light, so that the creature possessed great powers of vision.

The organic remains associated with fish-lizards tell us that they inhabited waters of moderate depth, such as prevails near a coast-line or among coral islands. Moreover, an air-breathing creature would obviously be unable to live in "the depths of the sea;" for it would take a long time to get to the surface for a fresh supply of air.

Perhaps no part of the skeleton is more interesting than the curious circular series of bony plates surrounding the iris and pupil of the eye. The eyes of many fishes are defended by a bony covering consisting of two pieces; but a circle of bony overlapping plates is now only found in the eyes of turtles, tortoises, lizards, and birds, and some alligators. This elaborate apparatus must have been of some special use; the question is—What service or services did it perform? Here, again, we find answers suggested by Owen and Buckland. It would aid, they say, in protecting the eye-ball from the waves of the sea when the creature rose to the surface, as well as from the pressure of the water when it dived down to the bottom—for even at a slight depth pressure increases, as divers know.

It might well be supposed that no record had been preserved from which we could learn anything about the nature of the skin of our fish-lizard; but even this wish has been partly fulfilled, to the delight of all geologists. Certain specimens have been obtained, from the Lias of England and Germany, that show faithful impressions of the skin that covered the paddles. A

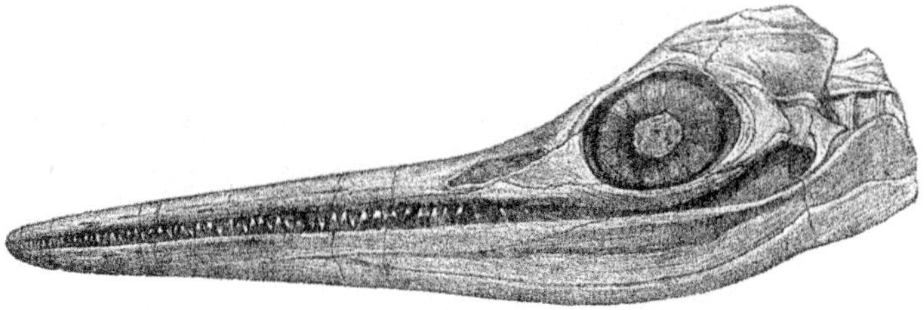

FIG. 11.—Head of *Ichthyosaurus platyodon*.

specimen of this nature has been presented to the national treasure-house at South Kensington by Mr. Montagu Browne.

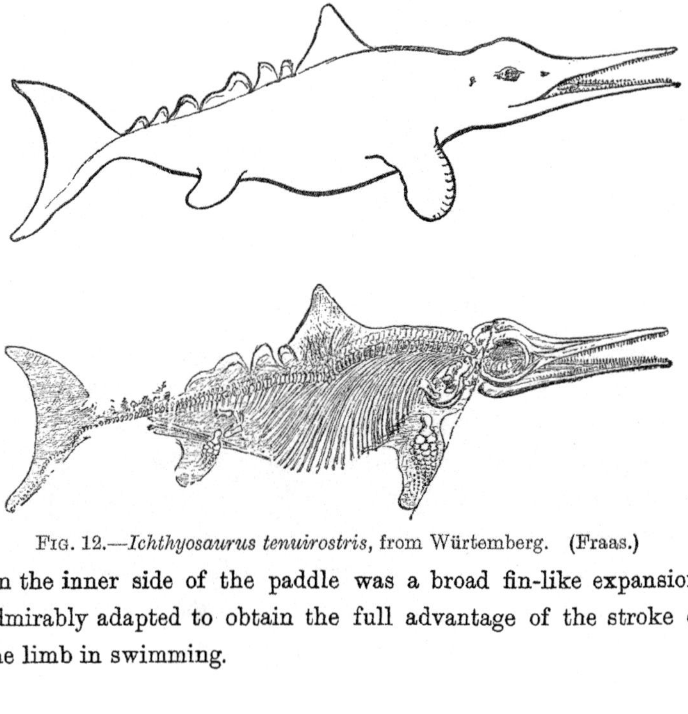

FIG. 12.—*Ichthyosaurus tenuirostris*, from Würtemberg. (Fraas.)

On the inner side of the paddle was a broad fin-like expansion, admirably adapted to obtain the full advantage of the stroke of the limb in swimming.

In the year 1892 there was discovered in the Lias of Würtemberg the skeleton of an Ichthyosaur, in which the outline of the fleshy parts is completely preserved (see lower figure). The reader will see from the figure that the tail-fin is very large, and the backbone appears to run into the lower lobe. Such a tail-fin as this impression indicates must have resembled that of the shark, only it is wider; but the shark's backbone runs into the *upper* lobe. Sir Richard Owen long ago foretold the existence of this appendage, and the discovery, coming now, adds one more tribute to his genius.

As Dr. Fraas remarks, this discovery shows how closely analogous Ichthyosaurs were in form to fishes, and further justifies the title of "fish-lizards." He considers that they did *not* visit the shore.

Speaking of the limbs, it should be mentioned that the bones of each finger, instead of being elongated and limited in number to three in each of the five fingers, are polygonal in shape and arranged in as many as seven or eight rows, while those of each finger are exceedingly numerous. Thus the whole structure forms a kind of bony pavement which must have been very supple. Such a limb would be one of the most efficient and powerful swimming organs known in the whole animal kingdom. In whales the fingers of the flippers are of the usual number, namely, five. Some species of fish-lizards had as many as over a hundred separate little bones in the fore paddle.

Another question naturally suggests itself: Were they viviparous, or did they lay eggs like crocodiles? This question seems to have been answered in favour of the first supposition; and in the following interesting manner. It not infrequently happens that entire little skeletons of very small individuals are found under the large ones. They are invariably uninjured, and of the same species as the one that encloses them, and with the head pointing in one direction. Such specimens are most probably the fossilised remains of little fish-lizards, that were yet unborn when their mothers met with an untimely end. In some cases, however, they may be young ones that were swallowed.

The jaws of these hungry formidable monsters were provided with a series of formidable teeth—sometimes over two hundred in number—inserted in a long groove, and not in distinct sockets, as in the case of crocodiles. In some cases, sixty or more have been found on each side of the upper and lower jaws, giving a total of over two hundred and forty teeth! The larger teeth may be two inches or more in length.

The jaws were admirably constructed on a plan that combined lightness, elasticity, and strength. Instead of consisting of one piece only, they show a union of plates of bone, as in recent crocodiles. These plates are strongest and most numerous just where the greatest strength was wanted, and thinner and fewer towards the extremities of the jaw. A crocodile, Sir Samuel Baker says, in his *Wild Beasts and their Ways*, can bite a man in two; and no doubt our fish-lizard would have been glad to perform the same feat! But in his pre-Adamite days the opportunity did not present itself.

The spinal column, or backbone, with its generally concave vertebræ, must have been highly flexible, as is that of a fish, especially the long tail which the creature worked rapidly from side to side as it lashed the waters.

Our description of the fish-lizard has, we trust, been sufficient—although not couched in the language used by men of science—to give a fair idea of its structure and habits.

In conclusion, a few words may be said about the ancestry and life-history of these ancient monsters. Palæontologists have good reason to believe that they were descended from some early form of land reptile. If so, they show that whales are not the first land animals that have gone back to the sea, from which so many forms of life have taken their rise.

During the long Mesozoic period fish-lizards played the part that whales now play in the economy of the world; and they resembled the latter, not only in general shape, but in the situation of the nostrils (near the eye), and in their teeth and long jaws. But these curious resemblances must not be interpreted to mean that whales and fish-lizards are related to each other. They only show that similar modes of life tend to produce artificial resemblances—just as some whales, in their turn, show a superficial resemblance to fishes (see p. 68).

With regard to the particular form of reptile from which the fish-lizard may have been derived, no certain conclusion has at present been arrived at. This is chiefly from want of fuller knowledge of early forms, such as may have existed in the previous periods known as the Carboniferous and Trias (see Appendix I.). But there are certain features in the skulls, teeth, and vertebræ that suggest a relationship with the Labyrinthodonts, such as Archægosaurus (see p. 96), or primæval salamanders that flourished during the above periods, or at least from amphibians more or less closely allied to them. They cannot by any possibility be regarded as modified fishes; for fishes have gills instead of lungs. Professor Merriam has found in the Triassic strata of California some of the links between some old land-reptiles and the fish-lizards; but at present no early marine ancestors have come to light.

The fish-lizards played their part, and played it admirably; but their days were numbered, and the place they occupied has since been taken by a higher type—the mammal. As reptiles, they were eminently a success; but, then, they were only reptiles, and therefore were at last left behind in the struggle for existence, until finally they died out, at the end of the Cretaceous period, when certain important geographical and other changes took place, helping to cause the extinction of many other strange forms of life, as we shall see later on.

They had a wide geographical range; for their remains have been discovered in Arctic regions, in Europe, India, Ceram, North America, the east coast of Africa, Australia, New Zealand, and Chili.

In American deposits they are represented by certain toothless forms, to which the names Sauranodon ("toothless lizard") and Baptanodon have been given. These have been discovered by Professor Marsh, in the Jurassic strata of the Rocky Mountains. They were eight or nine feet long, and in every other respect resembled Ichthyosaurs. As we have endeavoured to indicate in our illustration, the fish-lizards flourished in seas wherein animal, and doubtless vegetable life was very abundant. Any one who has collected fossils from the Lias of England will have found how full it is of beautiful organic remains, such as corals, mollusca, encrinites, sea-urchins, and other echinoids, fishes, etc.

The climate of this period in Europe was mild and genial, or even semi-tropical. Coral reefs and coral islands varied the landscape. There is just one more point of interest that ought not to be omitted; it refers to the manner in which these reptiles of the Lias age met their deaths, and were thus buried up in their rocky tombs. Sir Charles Lyell and other writers point out that the individuals found in those strata must have met with a sudden death and quick burial; for if their uncovered bodies had been left, even for a few hours, exposed to putrification and the attacks of fishes at the bottom of the sea, we should not now find their remains so completely preserved that often scarcely a single bone has been moved from its right place. What was the exact nature of this operation is at present a matter of doubt.[1]

[1] Just before going to press comes news of another most interesting find near Würtemberg of a small Ichthyosaur showing the whole of the skin on one side. See *Illustrated London News*, Jan. 1st, 1910.

CHAPTER V

THE GREAT SEA-LIZARDS AND THEIR ALLIES

"The wonders of geology exercise every faculty of the mind—reason, memory, imagination; and though we cannot put our fossils to the question, it is something to be so aroused as to be made to put the question to one's self."—HUGH MILLER.

THE fish-lizards, described in our last chapter, were not the only predaceous monsters that haunted the seas of the great Mesozoic age, or era. We must now say a few words about certain contemporary creatures that shared with them the spoils of those old seas, so teeming with life. And first among these—as being more fully known—come the long-necked sea-lizards, or Plesiosaurs.

The Plesiosaurus was first discovered in the Lias rocks of Lyme Regis, in the year 1821. It was christened by the above name, and introduced to the scientific world by the Rev. Mr. Conybeare (afterwards Dean of Llandaff) and Mr. (afterwards Sir Henry) de la Beche. They gave it this name in order to distinguish it from the Ichthyosaurus, and to record the fact that it was more nearly allied to the lizard than the latter.[1] Conybeare, with the assistance of De la Beche, first described it in a now-classic paper read before the Geological Society of London, and published in the *Transactions* of that Society in the year

[1] The name is derived from two Greeks words—*plesios*, near, or allied to, and *sauros*, a lizard.

1821. In a later paper (1824) he gave a restoration of the entire skeleton of Plesiosaurus dolichodeirus; and the accuracy of that restoration is still universally acknowledged. This fine specimen was in the possession of the Duke of Buckingham, who kindly placed it at the disposal of Dr. Buckland, for a time, that it might be properly described and investigated.

A glance at our illustration, Plate VI., will show that this strange creature was not inaptly compared at the time to a snake threaded through the body of a turtle.

Perhaps the best way in which we can gain a clear idea of the general characters of a long-necked sea-lizard, as we may call our Plesiosaurus, is by comparing it with the fish-lizard, described in the last chapter. Its long neck and small head are the most conspicuous features. Then we notice the short tail. But if we compare the paddles of these two extinct forms of life, we notice at once certain important differences. In the fish-lizard the bone of the arm was very short, while all the bones of the fore arm and fingers were modified into little many-sided bodies, and so articulated together as to make the whole limb, or paddle, a solid yet flexible structure. In the long-necked sea-lizard, however, we find a long arm-bone with a club-like shape; and the two bones of the fore arm are seen to be longer than in the fish-lizard. But a still greater difference shows itself in the bones of the finger, as we look at a fossilised skeleton (or a drawing of one); for the fingers are long and slender, like those of ordinary reptiles.

There are only five fingers, and each finger is quite distinct from the others. This is the reason why the Plesiosaur was considered to depart less from the type of an ordinary reptile, and so received its name. Other remarkable differences present themselves in the shoulders and haunches, but these need not be considered here. The species shown in Fig. 14 had rather a large head. It is obvious that such a long slender neck as these creatures had could not have supported a large head, like that of the fish-lizard. Consequently, we find a striking contrast in the skulls of the two forms. That of the Plesiosaur was short

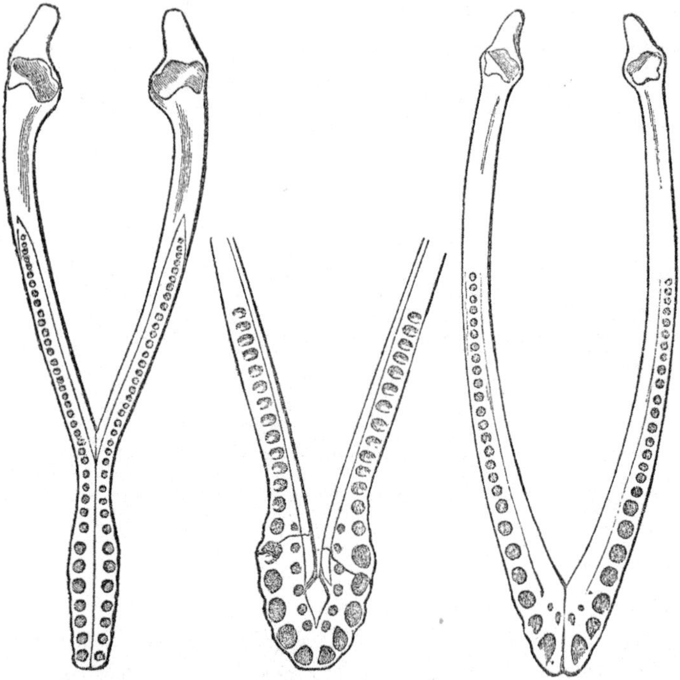

A B C

FIG. 13.—Mandibles of Sea-lizards. A, *Peloneustes philarchus* (Seeley); from the Oxford Clay. B, *Thaumatosaurus indicus* (Lydekker); Upper Jurassic of India. C, *Plesiosaurus dolichodirus* (Conybeare); from the Lower Lias, Lyme Regis.

and stout, and therefore such as could easily be supported, as well as rapidly moved about by the long slender neck. Thus we find another simple illustration of the "law of correlation," alluded to on p. 7. The teeth were set in distinct sockets, as they are in crocodiles, to which animals there are also points

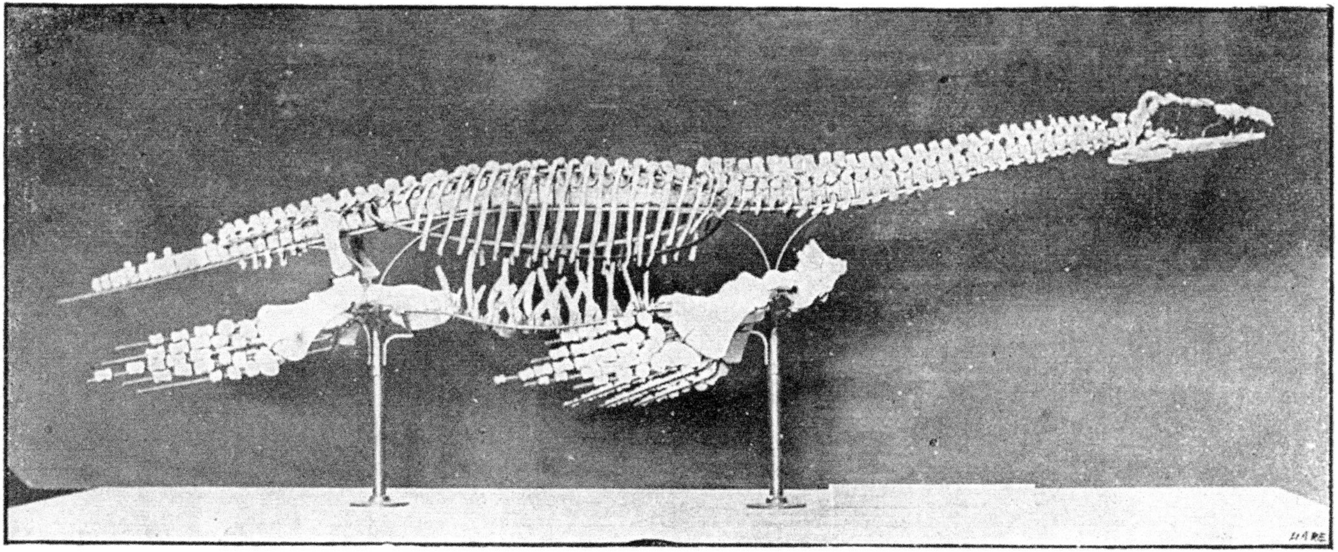

Plate V.

SKELETON OF YOUNG PLESIOSAUR, FROM OXFORD CLAY, NEAR PETERBOROUGH.

Natural History Museum, London.

of resemblance, in the backbone, ribs, and skull. Fig. 13 shows three different types of lower jaws of Plesiosaurs. The one marked C belongs to Plesiosaurus dolichodirus, the species represented in our plate. There were no bony plates in the eye. Professor Owen thinks that they were long-lived. The skin was probably smooth, like that of a porpoise.

The visitor to the geological collection at South Kensington will find a splendid series of the fossilised remains of long-necked sea-lizards. They were mostly obtained from the Lias formation of Street in Somersetshire, Lyme Regis in Dorset, and Whitby in Yorkshire. Those from the Lias are mostly small, about eight to ten feet in length. But in the rocks of the Cretaceous period, which was later, are found larger specimens. There is a cast of a very fine specimen from the Upper Lias on the wall of the Geological Gallery No. IV., at South Kensington, which is twenty-two feet long. But some of the Cretaceous forms, both in Europe and America, attained a length of forty feet, and had vertebræ six inches in diameter. The bodies of the vertebræ, or "cup-bones," are either flat or slightly concave, showing that the backbone as a whole was less flexible than in the fish-lizards.

Plate V. is from a photograph of a Plesiosaurus skeleton set up as if it were that of a living animal in the Natural History Museum, in 1892, the bones having been carefully collected by Mr. Alfred N. Leeds, of Eyebury, who has disinterred the separate bones of many distinct skeletons of Plesiosaurs from Oxford Clay strata near Peterborough, and to whom great credit is due for his laborious work. Quite recently another big Plesiosaur skeleton has been mounted in the Natural History Museum, near to the above. Though much disturbed by crushing, it shows some remarkable and novel features.

Specimens of extinct animals thus set up, give one a much better idea than when the bones are all lying huddled together on a slab of rock. But it is not always possible to get the bones entirely out of their rocky bed, or matrix.

FIG. 14.—*Plesiosaurus macrocephalus.*

It will be remembered that the long and powerful tail of the fish-lizard was its principal organ of propulsion through the water; and that, consequently, the paddles only played a secondary part. They were small, but amply large enough for the work they had to perform. But our long-necked sea-lizards possessed very short tails. What, then, was the consequence? Obviously that the paddles had all the more work to do. They were the chief swimming organs. The vertebræ of this short tail show that it probably was highly flexible, and could move rapidly from side to side; but, for all that, its use as a propeller would not be of much importance. We see now why the paddles are so long and powerful, like two pairs of great oars, one pair on each side of the body. In a fossil skeleton you will notice the flattened shape of the arm-bone (or humerus), and of the thigh-bone (or femur). This gave breadth to the paddles, and made them more efficient as swimming organs. They give no indication of having carried even such imperfect claws as those of turtles and seals, and therefore we may conclude that the Plesiosaur was far more at home in the water than on land, and it seems probable that progression on land was impossible.

The tail was probably useful as a rudder, to steer the animal when swimming on the surface, and to elevate or depress it in ascending and descending through the water. Like the fish-lizard, this creature was an air-breather, and therefore was obliged occasionally to visit the surface for fresh supplies of air. But probably it possessed the power of compressing air within its lungs, so that the frequency of its visits to the surface would not be very great.

From the long neck and head, situated so far away from the paddles, as well as for other reasons, it may be concluded that this creature was a rapid swimmer, as was the Ichthyosaurus. Although of considerable size, it probably had to seek its food, as well as its safety, chiefly by artifice and concealment. The fish-lizard, its contemporary, must have been a formidable rival and a dangerous enemy, whom to attack would be unadvisable.

Speaking of the habits of the long-necked sea-lizard, Mr. Conybeare, in his second paper, already alluded to, says, "That it was aquatic, is evident from the form of its paddles; that it was marine, is almost equally so, from the remains with which it is universally associated; that it may occasionally have visited the shore, the resemblance of its extremities to those of the turtle may lead us to conjecture; its motion, however, must have been very awkward on land; its long neck must have impeded its progress through the water, presenting a striking contrast to the organisation which so admirably fits the Ichthyosaurus to cut through the waves.

"May it not therefore be concluded (since, in addition to these circumstances, its respiration must have required frequent access of air) that it swam upon or near the surface, arching back its long neck like the swan, occasionally darting it down at the fish which happened to float within its reach? It may, perhaps, have lurked in shoal-water along the coast, concealed among the sea-weed, and, raising its nostrils to a level with the surface from a considerable depth, may have found a secure retreat from the assaults of dangerous enemies; while the length and flexibility of its neck may have compensated for the want of strength in its jaws and its incapacity for swift motion through the water, by the suddenness and agility of the attack which they enabled it to make on every animal fitted for its prey, which came within its extensive sweep."

More than twenty species of long-necked sea-lizards are known to geologists.

Professor Owen, in his great work on *British Fossil Reptiles*, when describing the huge Plesiosaurus dolichodirus from Dorset, suggests that the carcase of this monster, after it sank to the bottom of the sea, was preyed upon by some carnivorous animal (perhaps sharks). It seems, he says, as if a bite of the neck had pulled out of place the eighth to the twelfth vertebræ. Those at the base of the neck are scattered and dispersed as if through

PTERODACTYLS. LONG-NECKED SEA-LIZARD. CUTTLE-FISH.

PLATE VI. *Plesiosaurus dolichodeirus.* Length 22 feet.

more "tugging and riving." So with regard to its body, probably some hungry creature had a grip of the spine near the middle of the back, and pulled all the succeeding vertebræ in the region of the hind limbs. Thus we get a little glimpse of scenes of violence that took place at the bottom of the bright sunny seas of the period when the clays and limestones of the Lias rocks were being deposited in the region of Lyme Regis.

As time went on, these curious reptiles increased in size, until, in the period when our English chalk was being formed (Cretaceous period), they reached their highest point (see p. 197). After that they became extinct—whether slowly or somewhat suddenly we cannot tell.

Until more is known of the ancient life of the earth, it will not be possible to say with certainty what were the nearest relations of the long-necked sea-lizards. They first appear in the strata of the New Red Sandstone, which is below the Lias. Certain little reptiles, about three feet long, from the former rocks, known as Neusticosaurus and Lariosaurus, seem to be rather closely related to the creatures we are now considering, and to connect them with a group of land reptiles. They were partly terrestrial and partly aquatic; but it is not easy to say whether their limbs had been converted into true paddles or not. At any rate, there is every reason to believe that the long-necked sea-lizards were descended from an earlier form of land reptile. They gradually underwent considerable modifications, in order to adapt themselves to an aquatic life. We noticed that the same conclusion had been arrived at with regard to the fish-lizards. Both these extinct groups, therefore, present an interesting analogy to whales, which are now considered to have been derived, by a like series of changes, from mammals that once walked the earth.

The Plesiosaur presents, on the one hand, points of resemblance to turtles and lizards,—on the other hand, to crocodiles, whales, and, according to some authorities, even the strange Ornithorhynchus. But it will be very long before its ancestry can be made known. In the mean time we must put it in a place somewhere near the fish-lizards, and leave posterity to complete what has at present only been begun. It must, however, be borne in mind that some of the above resemblances are purely accidental, and not such as point to relationship. Because their flippers are like those of a whale, it does not mean that Plesiosaurs are related to modern whales. It only means that similar habits tend to produce accidental resemblances—just as the whales and porpoises, in their turn, resemble fishes. To make torpedoes go rapidly through the water, inventors have given them a fish-like shape;—in the same way the early forms of mammals, from which whales are descended, gradually adapted themselves to a life in the water, and so became modified to some extent to the shapes of fishes.

The Pliosaurs, above mentioned, are evidently relations, but with short necks instead of long ones. They had enormous heads and thick necks. Fine specimens of their huge jaws, paddle-bones, etc., may be seen in the fossil reptile-gallery at Cromwell Road. One of the skulls exhibited there is nearly six feet long, while a hind paddle measures upwards of six and a half feet in length, of which thirty-seven inches is taken up by the thigh-bone alone. The teeth at the end of the jaws are truly enormous. One tooth, from a deposit known as the Kimmeridge Clay, is nearly a foot long from the tip of the crown to the base of the root. In some, the two jaw-bones of the lower jaw are partly united, as in the sperm-whale or cachalot. Creatures so armed must have been very destructive.

CHAPTER VI

ANCIENT SALAMANDERS

"Slowly moves the march of ages,
Slowly grows the forest-king,
Slowly to perfection cometh
Every great and glorious thing."

ANON.

ONE of the great steps in the upward progress of the animal kingdom was the transition from the fish to the reptile. These two classes are at the present time well marked and clearly separated from each other; but an idea of how the change took place, in the course of Evolution, may be gathered from a study of the important group, or class, which comes in between the two, viz. the amphibians, or batrachia of some naturalists. This important and interesting group forms, as it were, a series of "links" between fishes and reptiles. They derive the latter name from the Greek word *batrachos*, signifying a frog. But we will speak of them as amphibia,[1] because they live both in water and on land; and during the early part of their lives swim in water like fishes, and moreover breathe as fishes do by means of gills. Afterwards they put away such childish things as gills, and, developing lungs, become air-breathers, and behave like ordinary reptiles.

Modern naturalists who accept the theory of Evolution, as most

[1] Greek—*amphi*, both; and *bios*, life.

do, regard the metamorphosis of the tadpole into a frog as a kind of object-lesson on the part of Dame Nature, given with a view to show us, in a general way and in a short space of time, how the wonderful change from the fish to the reptile took place.

By the time the great coal-forests of the Carboniferous period were flourishing, the fish class had attained a high state of development, and were to some extent foreshadowing the amphibian type of creation. Now, we know that when the New Red Sandstone was being laid down, the next great class, viz. the reptiles, had not only made a start, but had struck out in several lines, and were specialised in certain directions, as shown by their bird-like footprints on the Triassic sandstones. Would it not therefore be expected that we should find, in rocks of greater age than the New Red Sandstone, the remains of animals in some respects intermediate between fishes and reptiles? This expectation is fully realised by the presence in such strata (*i.e.* the coal-measures and Permian rocks) of the remains of quite a large number of amphibia. They also crop up again in the Trias, but after that period are not much in evidence until we come to Tertiary times.

The amphibian class, at the present day, is represented by the frogs, toads, and newts of our own country, and by the salamanders of Japan and the United States. But these are only a meagre remnant of a large number of families that lived ages ago.

Let us endeavour now to take a general view of the amphibian class as it was in the days of long ago, before the creation of man,—to look at their family tree and see what their ancestors were like. When this is done we shall perhaps have a little more respect for the frogs and newts that inhabit our ponds; for they come of a very ancient stock, and one which played an important part in primæval times, branching out here and there into strange and varied forms, as if they could not help themselves. In order to do this we must go a long way back into the world's history; and even then we shall still be some way off the roots of the family tree; or, in other words, we shall be in the dark about the first forms of amphibian life which appeared on the earth's surface.

To find the oldest known amphibian we must go back as far as those "dark ages" when a large portion of the surface of what is now Europe was overgrown with vast areas of dismal coal-forests, something like the mangrove swamps of which we read in books of African travel; they must have been dark compared to an English meadow on a summer day, and therefore the expression "dark ages" is all the more appropriate. Our knowledge of the earliest air-breathing creatures is no doubt small, but of late years it has been considerably extended, thanks to the labours of several English, American, and Continental workers, such as Owen, Huxley, Miall, Seeley, Gaudry, Fritsch, Cope, and others, who have all carefully studied and described the fossil remains submitted to them, to say nothing of the numerous eager collectors who have brought away from the bowels of the earth numerous specimens of great value. Their services, although of a more humble order, are quite indispensable; and it should be borne in mind that it is in the power of all of us, however ignorant we may be of comparative anatomy, and therefore unable to write learned papers for the Geological Society, to render valuable service to the cause of Science simply by using our eyes and not neglecting opportunities of collecting specimens wherever we may go.

But before proceeding it will be necessary to introduce a new name for our old amphibians. It is rather a long one, we admit, but useful, because it partly serves to describe them. Scientific men are said to be very fond of long Latin and Greek names, but such can hardly be avoided; and there is this to be said for them—that the names they use are much more exact than the popular ones by which plants and animals are commonly known, and, moreover, they often convey a great deal of information in a brief manner. Now, the name given to the creatures we are about to describe is Labyrinthodonts.

It was Sir Richard Owen who first accurately described the fossil amphibians, and they received this name from him. Let us see what the name means and how it applies. Sir Richard Owen's acquaintance with these remarkable forms of ancient life began in the year 1840, when he first examined certain teeth from the New Red Sandstone of Coton End quarry, Warwickshire. In external character these fragmentary teeth corresponded with those which had been previously discovered in Germany (in Würtemberg) by Professor Jaeger, and which had been called by him Mastodonsaurus. This name was not a happy one, because it suggested an association with an extinct elephant, the Mastodon, with which this amphibian, it is hardly necessary to say, had no connection, real or imaginary. Owen examined the teeth of the fossils from Germany, as well as those from Warwickshire, and found that, when cut across into transverse sections for the microscope, they revealed a very remarkable and complicated structure; the whole of the internal portion was seen to be made up of a complex series of foldings, forming a peculiar structure, suggesting a labyrinth—the external layer of cement belonging to the tooth converging in numerous folds towards the pulp-cavity. And so the name Labyrinthodonts reminds us that all, or nearly all, the fossil animals included under the above general term possess teeth having this peculiarity. They are the Labyrinth-toothed amphibians (see Fig. 15).

At first sight it might perhaps seem to the uninitiated as if the internal and minute structure of a tooth were a matter of comparative unimportance; but it is recognised by all anatomists that teeth are highly important as indicating the place of an animal in the scale of being. In the present case, for instance, the possession of teeth with the labyrinthine structure indicates an affinity with certain ganoid fishes, which also possess similarly infolded teeth. Again, there is a peculiarly low type of reptiles, known as the Ichthyosaurus, or "fish-lizard," of which the teeth show an approach to the same kind of structure. Some of the Labyrinthodonts retained the gills of their youth through life, instead of changing them for lungs, as frogs do.

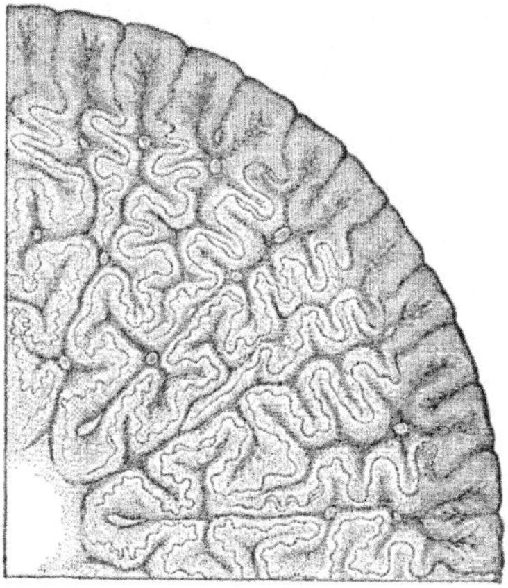

FIG. 15.—Magnified section of tooth of *Labyrinthodont.* (After Owen.)

But, if these facts tend to connect the Labyrinthodonts with fishes below, there are others which tend to connect them with the reptiles above. Sir Richard Owen, by a bold piece of reasoning, arrived at the conclusion that certain huge and somewhat frog-like skulls, previously discovered in German Triassic rocks, belonged to Labyrinthodonts. This was an important step; but then the question arose, What kind of creature made the tracks found in these strata, both in Germany and England? This question cannot be answered with any certainty, as we have stated on p. 41. But we know that Anomodont reptiles and Dinosaurs abounded.

Like the reptiles of to-day, modern amphibians have no fixed type of external form. Some suggest a likeness to turtles and tortoises; others suggest in shape the true snakes; another group, represented by newts and salamanders, suggests a resemblance to the lizards; and lastly, those which keep their gills through life suggest certain finless fishes, such as the conger-eels and the lampreys. It was just the same in the old days, when amphibians flourished more abundantly than now; they struck out in many directions—some taking on a long snakelike form (see Plate VII. and Fig. 23), others being short and with larger limbs, while in the matter of size there were great differences.

As far as present knowledge goes, it seems to indicate that the reptilian class in early days was less advanced than in the later phases of the Secondary era (*i.e.* during the Jurassic and Cretaceous periods); and so we can understand that amphibians to some extent played the part of reptiles in those remote ages of the coal-forests and the New Red Sandstone. Their very diversity shows that they were in a flourishing condition, and probably had a pretty wide field to themselves, without too many enemies. A certain amount of competition with other creatures, no doubt, they had to submit to; but with fishes they could hold their own. Later on came a much more severe form of competition from highly developed reptiles, such as Dinosaurs, and that was too much for them (according to Darwinian teaching). As the reptile class rose, amphibians, of course, came off second best, and the geological record tells us that, ever since, they have remained in the background of the theatre of life.

It will be well, before proceeding to describe some of the better-known types of Labyrinthodonts, to say a word about the general characters by which they may be distinguished. Of the internal structure of the teeth we have already spoken ; but although the same complexity of folding does not apply to some of the smaller forms, the name by which they are known is very appropriate to the group as a whole. In Plate IX. the reader will find a restoration of a typical labyrinthodont the Mastodonsaurus, of which the skull is shown in Fig. 16. There was a marked difference in size between the hind limbs and fore limbs, as shown in the above Plate. There are five digits to each limb, and no claws.

Turning to their skulls, there is one feature that is quite peculiar to the group and rarely seen in reptiles, living or extinct, viz. that the whole of the upper surface of the skull behind the eye-holes is covered in by a complete roof of bone. Hence some authorities call them Stegocephali,[1] or "roof-skulled." This is clearly shown in Fig. 16, representing the huge skull of Mastodonsaurus. Some fishes have skulls thus roofed over. But the skull of a lizard or a crocodile is very different, showing a long open channel behind each orbit, and another one lower down at the side of the head. Another important character in the labyrinthodont skull is the presence of two condyles, instead of only one, as in all true reptiles and birds. The word condyle[2] signifies a knuckle, and is applied chiefly to the surface or surfaces by which the skull joins on to, or articulates with, the first vertebra of the neck. All the mammalia have skulls articulating by two condyles—a fact which would seem to imply that mammals branch off from the amphibia, and not from reptiles. Again, the bones of which the skull is composed are nearly always covered with a network of grooves, or "canals," doubtless intended for the attachment of hard, horny, and bony scales or scutes.

[1] Greek—*stegos*, roof; *cephalos*, head.

[2] Greek—*condulos*.

Before we leave the skull there is one other very interesting point that should not be omitted. If the reader will refer to the skull of Mastodonsaurus (Fig. 16), he will see, just on the middle line and near the base, a small round hole. It lies over the brain-cavity, and most probably represents the site of a small third eye. Whether this little extra eye was capable of receiving impressions of light, and so of communicating them to the brain

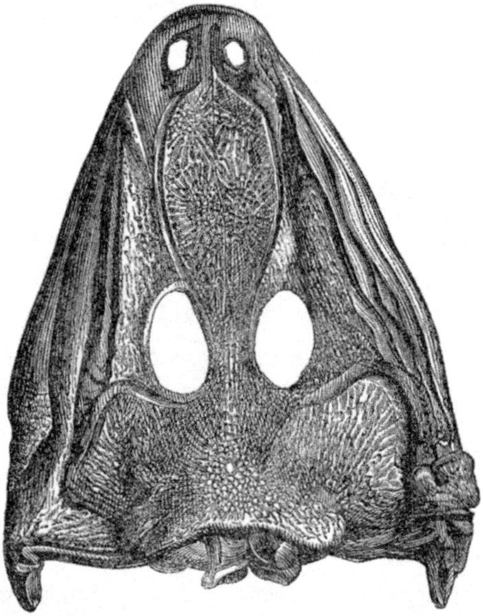

FIG. 16.—Skull of a Labyrinthodont *Mastodonsaurus Jaegeri*, from New Red Sandstone, Germany.

as an ordinary eye does, is a matter of doubt; but in a certain very primitive little lizard now living in New Zealand, and known as the Tuatara (Sphenodon), there is a similar aperture overlying the rudiment of an eye now quite useless and lying down in the brain. This little eye in the Tuatara must at one time have been made use of, and the question arises whether the third eye of the Labyrinthodonts was similarly useless; but the chances are rather against the notion. All the lizards have this pineal eye. It is probable that we have here a relic from some much older and now lost form of life—possibly a fish, but possibly also a still lower type. These rudimentary structures, of which Darwin speaks in his great work, *The Origin of Species*, always possess a high degree of interest for the palæontologist, as being survivals from very early days in the world's history. To the latter they are as interesting in their way as are ancient customs and old superstitions to a student of folk-lore or of ancient history.

We must also notice the forward position of the creature's nostrils as indicated by the two small openings at the apex of the skull (see Fig. 16). If we were to examine the under side of a skull such as that of the Mastodonsaurus, we should see that the bony palate is formed chiefly of two broad and flat bones, called the "vomerine," which generally support teeth. Now, this is a character not exhibited by the skull of any true reptile, and we must go to modern amphibia, such as our frogs and toads, to find a parallel. There is also a corresponding series of small teeth on the mandible of the lower jaw (see Fig. 25). The eyes often show the peculiar sclerotic plates seen so well in the "fish-lizards."

In studying an extinct vertebrate, or backboned animal, one of the first points to be considered is the nature of the vertebræ, or joints of the backbone. Now, in all fishes, and also in their distant cousin the Ichthyosaurus (fish-lizard), these bones have their articulating surfaces (centra) hollow, hence the popular term "cup bones" applied to the vertebræ of the latter. Need we be surprised, then, to learn that the Labyrinthodonts possessed similarly cup-shaped vertebræ? This is one more link to connect them with fishes. In some of the more primitive and ancient forms, the body of the vertebra was composed of three separate pieces, an arrangement peculiar to certain primitive fishes.

Ever since an unfortunate restoration of a Labyrinthodont by Mr. W. Hawkins, it has been the custom—in popular works at least—to represent the creature with a frog-like aspect. This restoration was based on the imperfect material of a good many years ago, and is therefore out of date. But later discoveries have shown that, instead of being frog-like, most of the Labyrinthodonts, if we could have seen them in the flesh, would have reminded us more of our modern little newts, or of the salamanders of hot countries. The tail was generally well developed, and the limbs were adapted quite as much for walking as for swimming.

Their chests were defended by three bony plates (see Fig. 17), which had grooves similar to those on the skull. But besides this, all the under side of the body was protected by a large number of scales, running in two directions, as shown in Fig. 17. In one or two instances the whole body was covered with a coat of mail, consisting of small bony scales. An example of this kind of skin is the Seeleya, from the Permian rocks of Bohemia, restored by Fritsch, and named by him after the late Professor H. G. Seeley, F.R.S. In size the Labyrinthodonts varied greatly, some being only an inch or two in length, others seven or eight feet, and perhaps more. Most of them had limbs, but in one or two cases these had apparently been dispensed with.

It might naturally be asked, Did they undergo a series of changes similar to the metamorphoses of a common frog? One could hardly, perhaps, have expected an answer to this question; but, strange to say, some of the smaller forms from the petroleum shales (Gaskohle) of Bohemia, so carefully studied and described

AMPHIBIANS OF THE COAL-FOREST AND PERMIAN PERIODS.

PLATE VII. *Actinodon.* *Ceraterpeton.* *Archegosaurus.* *Dolichosoma.* *Loxomma* (Head of).

by Dr. Fritsch, have been very well preserved in those deposits of the Permian age, so that he was able to recognise in

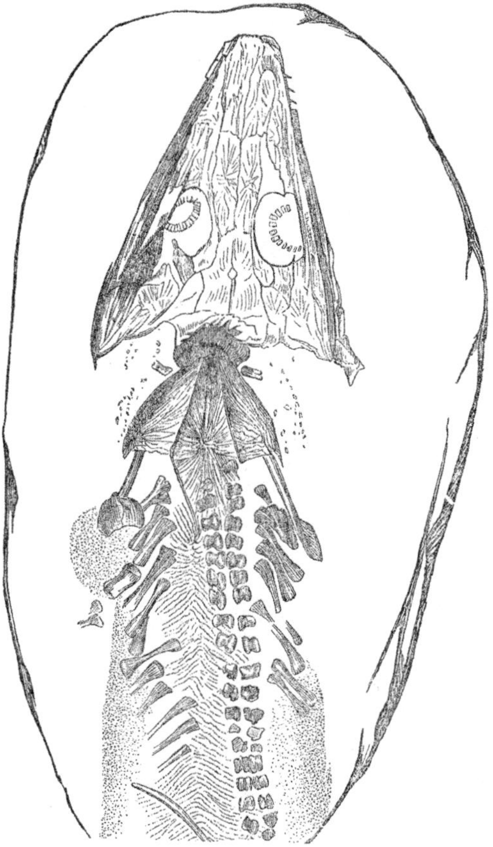

FIG. 17.—Part of skeleton of a Labyrinthodont, *Archegosaurus Decheni*, from the Permian, Saarbrücken. (After H. von Meyer.) The snout is incomplete. The three ventral plates are seen near the head.

some cases the signs of the gills, by means of which, in early life, they breathed in water, like fishes. And so it is manifest that they did go through a series of metamorphoses.

One of the most primitive, and at the same time one of the earliest known Labyrinthodonts, is the Archegosaurus, of which we must now give a short account.

It resembled the modern Proteus and other amphibians that retain their lungs throughout life.

Some of the specimens acquired since Sir R. Owen published his conclusions on this creature are now in the Natural History Museum. There are certain fishes of the present day that have their backbones in an incomplete or unfinished state; that is to say, the bones consist partly of cartilage, like an unfinished house which is only built up to a certain level, the upper stories being merely indicated by scaffolding. An example of this kind of fish is furnished by the existing mud-eel, or Lepidosiren, a highly interesting form of life to the geologist, because it tends to fill up the gap between the two distinct classes of Fishes and Reptiles. Now, in this respect, the old Carboniferous amphibian we are now considering resembles the mud-fish. The latter leads a partly amphibious life, burying itself in the mud during the hot season, and breathing air by means of its air-sack. It is believed by evolutionists that, in the course of ages, lungs were developed from the air-sacks of "ganoid" fishes. After detailing certain other characters, in which Archegosaurus resembles the higher ganoid fishes, Sir Richard Owen says, "All these characters point to one great natural group, peculiar for the extensive gradations of development, linking and blending together fishes and reptiles within the limits of such group." The salamander-like ganoids, Lepidosteus of North America and Polypterus of the Nile, are the most fish-like; the true Labyrinthodonts are the most reptilian or saurian of the group; Archegosaurus conducts the march of development from the fish proper to the labyrinthodont type; Lepidosiren conducts it to the batrachian type with permanent gills. Both tend to show the artificial nature of the arrangement by which, under most systems of classification, fishes are separated by a wide gap from reptiles and amphibians.

The Labyrinthodonts ranged through three geological periods —the Carboniferous, the Permian, and the Triassic. At the time when the coal-forests of Europe and America were flourishing, they were tenanted by a large number of Labyrinthodonts. These, and others of the succeeding Permian period, which may conveniently be taken together, have been specially studied by such authorities as Professors Huxley and Cope, Dr. Fritsch,

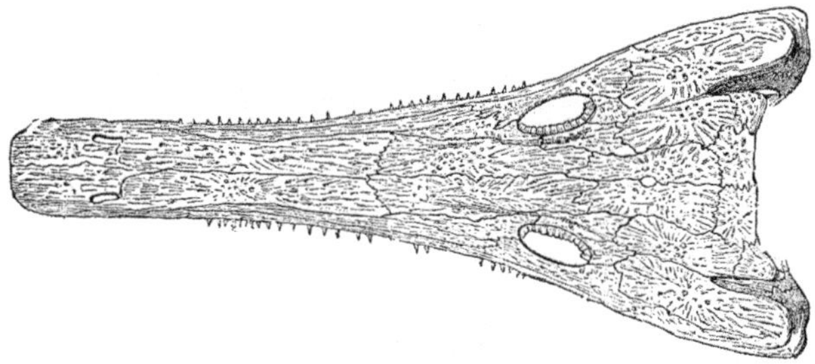

FIG. 18.—Skull of *Archegosaurus*. (Permian.)

Professor Miall, and others. The limits of space will only allow us to make a small selection from the numerous forms described by them.

Cricotus is an American genus described by Professor Cope, from the Permian strata of Texas. The skull and some of the scales belonging to the under surface of the body are shown in Fig. 19. The Eryops, of which the skull is shown in the same figure, is considered by Professor Seeley to belong to a higher, though related group, the Anomodonts (Chap. VII.). The one group is much connected with the other, and the question of the classification of their various members is a difficult one, which only time and further discoveries can settle.

Loxomma is another genus of the Coal period, but only the head is known (see Plate VII.). It was named by Professor Huxley, who, during a visit to Edinburgh in 1862, made a study of the large collection of vertebrate fossils from Burdie House and Gilmerton in the Museum of Edinburgh. While looking through the collection for fossil fishes, he came upon some specimens which were of quite a different character, and

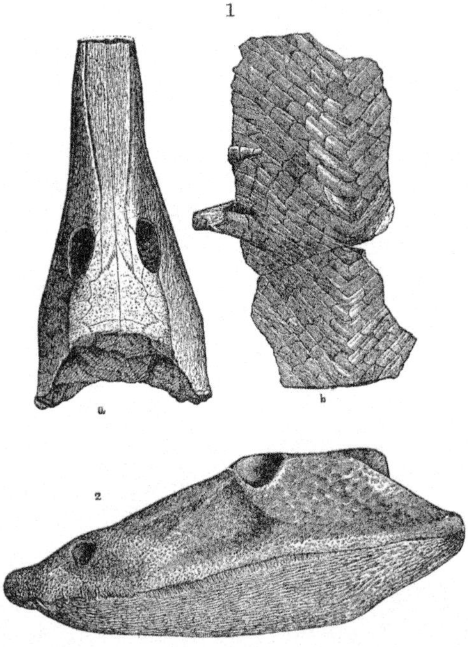

FIG. 19.—Labyrinthodonts. 1, *a*, *b*, *Cricotus*. 2. Skull of *Eryops*. From Permian strata, Texas. (After Cope.)

among these found part of the skull of a Labyrinthodont, which he named Loxomma. A very beautifully preserved specimen of the skull of this creature, from the coal-measures of Shropshire, is to be seen at the Natural History Museum (Gallery No. V. of the Geological Department). It is fortunately

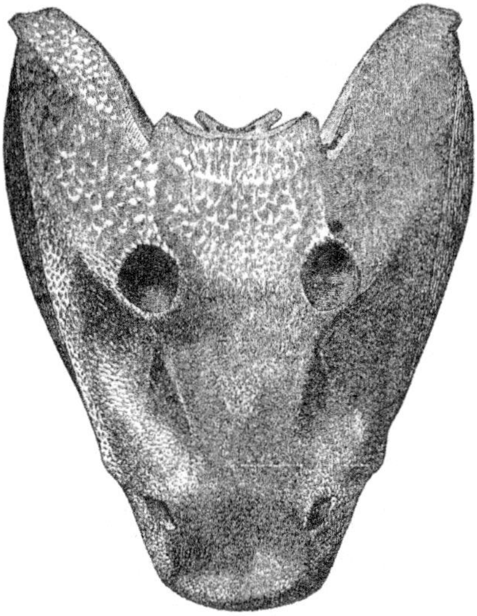

FIG. 20.—Skull of *Eryops macrocephalus*, from Permian strata, Texas. (After Cope.)

uncrushed, and shows the natural contour of the skull and lower jaw. The specimen was presented by George Maw, Esq., F.G.S.

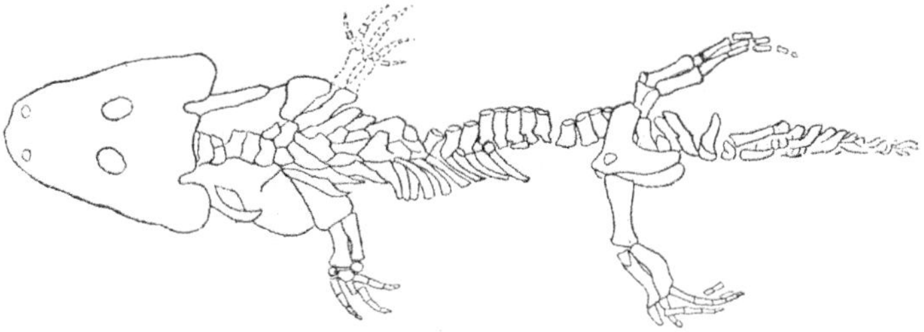

FIG. 21.—Skeleton of a Labyrinthodont, *Actinodon*, from Permian strata. (Partly after Gaudry.)

In Actinodon the skull is short and wide, the nostrils are large, and the muzzle is broad. Fig. 21 is a drawing of the complete skeleton, and in the wall-case 19, Gallery V., will be found a skull preserved on a slab of shale from Permian strata in

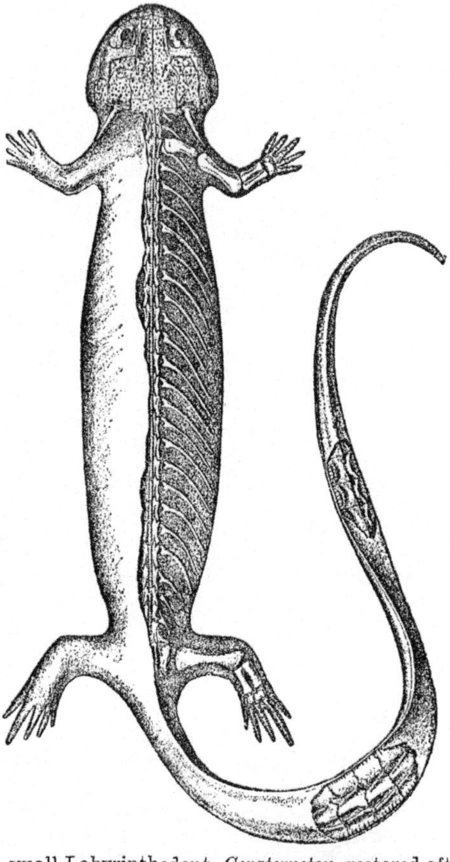

FIG. 22.—A small Labyrinthodont, *Ceraterpeton*, restored after Fritsch, from Permian strata, Bohemia.

France. A restoration of the creature when alive is shown in Plate VII.

A number of small Labyrinthodonts, somewhat like salamanders, have been found in Ireland and in Bohemia. Those

PTYONIUS (TWO SPECIES). (AFTER COPE.) FROM COAL MEASURES, OHIO.
PLATE VIII.

from the latter country are very exhaustively described by Dr. Fritsch, in a valuable monograph,[1] with many excellent illustrations. We have reproduced two of these, the Ceraterpeton and the snake-like Dolichosoma[2]—both restored in Plate VII., and their skeletons shown in Figs. 22, 23. Speaking of snake-like forms, it may be mentioned that one of them—the Palæosiren

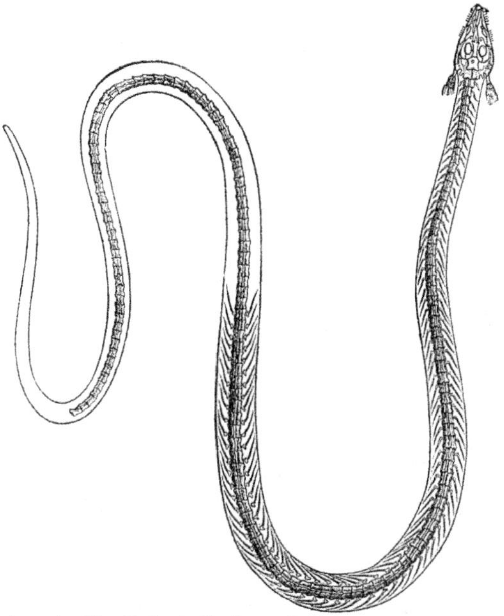

FIG. 23.—A snake-like Labyrinthodont, *Dolichosoma*, from Carboniferous strata, Bohemia. (After Fritsch.)

("ancient siren")—is estimated to have attained a length of forty-five feet. Professor Cope has written an account of a number of Labyrinthodonts from the coal-measures of Ohio, where both Professor Newberry and Professor Wyman have obtained specimens. These show great variety of form. Some resemble

[1] *Fauna der Gaskohle*, Dr. A. Fritsch.

[2] Greek—*dolichos*, long; *soma*, body.

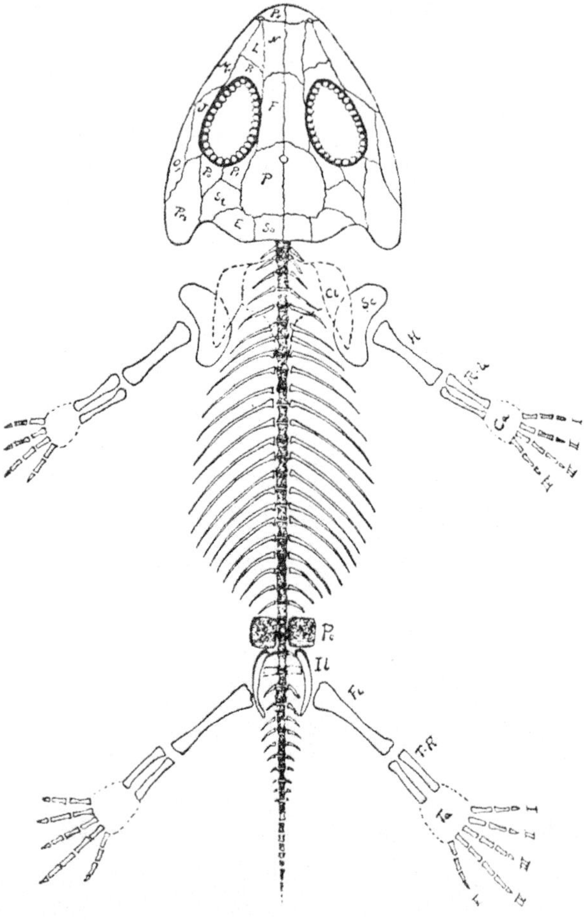

FIG. 24.—Restoration of skeleton of *Amphibamus grandiceps*, by R. L. Moodie. (*Journal of Geology*, vol. 17.)

long-limbed lizards; some have short limbs. Others, like Ptyonius (see Plate VIII.), were very elongate. The members of the Labyrinthodont order flourished over a large part of the earth's surface, from the time of the coal-forests to that of the Triassic sandstones. Their remains are found in all the great continents. The late Professor Seeley considered them to have been the ancestors of the crocodiles. Large skulls of Labyrinthodonts, such as Mastodonsaurus and Capitosaurus, have for a long time past been found in the Upper Triassic (Keuper) strata of Germany. They appear to resemble very closely a generalised Anomodont (see next Chapter).

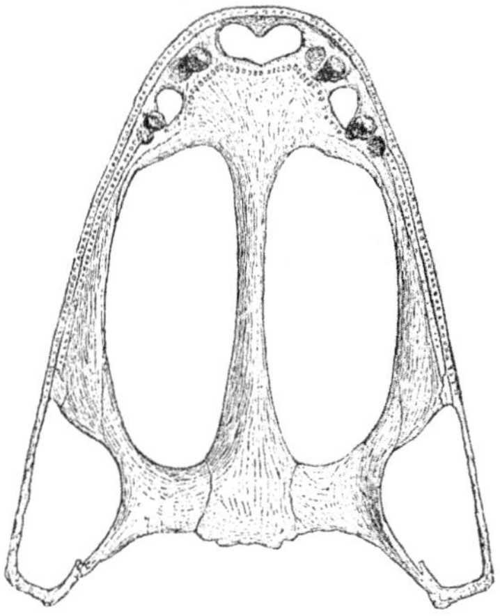

FIG. 25.—Skull of a Labyrinthodont, *Cyclotosaurus* (under side).

In the writer's previous work, *Creatures of Other Days*, p. 70, he reproduced a restoration of the skeleton of Labyrinthodon by Professor Wiedersheim, based on a discovery made in the year 1864 in Riehen, Switzerland. The specimen, now in the Museum at Basle, was somewhat fully described. Recent researches have shown that the interpretation given by Wiedersheim was wrong, and that the bones (rather badly preserved), represent a reptile of the Anomodont order, described in Chapter VII. This figure is now replaced by a recent restoration of the skeleton of another Labyrinthodont amphibian, viz. the Amphibamus of Mr. R. L. Moodie. Let the reader compare this with the skeleton of a true reptile, and he will at once see how primitive it is.

True Salamanders are known in a fossil state. There is the famous Andrias Scheuchzeri, about three feet long, found in 1725, and described by Scheuchzer as the skeleton of a man. He called it *Homo Diluvii Testis,* or, "the man who witnessed the Deluge." Cuvier, in 1811, announced its true nature.

REPTILES AND AN AMPHIBIAN OF THE NEW RED SANDSTONE PERIOD.

Rhynchosaurus. *Mastodonsaurus.* *Hyperodapedon.* *Telerpeton.*

PLATE IX.

CHAPTER VII

ANOMALOUS REPTILES

"In the endeavour to complete the Natural History of any class of animals, the mind seeks to penetrate the mystery of its origin, and, by tracing its mutations in time past, to comprehend more clearly its actual condition, and gain an insight into its probable destiny in time to come."—SIR R. OWEN.

HAVING, in our previous chapter, discussed the Labyrinthodonts, we now pass on to consider some of their descendants in the shape of a very peculiar group of reptiles hailing from South Africa and elsewhere. Perhaps the chief peculiarity, among many others, presented by these very antique and old-fashioned inhabitants of the world, is to be found in their teeth. In fact, so irregular and anomalous are they in regard to these organs, and so at variance with all our preconceived ideas with regard to what proper, well-behaved reptiles, whether living or extinct, ought to be like, that their vagaries in this respect have led to their being christened by the name of Anomodonts, or "anomalously toothed" reptiles.[1] Needless to say, they occupy a distinct order, with several sub-orders, in the classifications adopted by palæontologists. They are also by some authorities named Theromorphs and Theriodonts. The moral of all this is, as we shall better understand later on, that it would be well for students of extinct forms of life to enter this domain of Science without any preconceived ideas at all! It would save a great

[1] Greek—*anomos*, without law; *odous, odontos*, tooth.

deal of confusion and trouble in the end; and, moreover, would be far more truly scientific. For what right has any one, however great his knowledge or his ability, to dictate to Nature, and to say this or that is impossible—that no reptile, for instance, could possibly have flown; or that such and such teeth were impossible for a reptile?

We now know that there was a time when certain reptiles *did* fly (although many people with some pretensions to knowledge doubted the evidence). And so with regard to reptilian teeth; fossil evidence shows that some old reptiles had teeth more like those of modern mammals! Facts such as these should teach caution, and every student of palæontology will do well to remember the saying of Agassiz: "The possibilities of existence run so deeply into the extravagant that there is scarcely any conception too extraordinary for Nature to realise."

The chief characters of the Anomodonts may be briefly stated as follows: In this order the body is lizard-like, and the limbs are adapted for walking. The skull is comparatively short, and the nasal openings are large. The teeth are generally placed in distinct sockets (thecodont). The bodies of the vertebræ are hollow at both ends (amphicœlous), and in some cases are only partly converted into bone—a character which is common to the Labyrinthodonts. The whole structure of the foot is distinctly on the mammalian plan. Recent researches show that these animals are descended from Labyrinthodonts, and more especially the family of which Archegosaurus is a member (see p. 96). Certain important characters show (strange as it may seem) affinity with mammals; and it is probable that they are related to the lowest group of them, as represented at the present day by that remarkable creature the Spiny Ant-eater of Australia (Echidna), and the wonderful Duck-mole (Ornithorhynchus), which lays eggs like a reptile. These two creatures belong to the Monotreme order.

Now the Anomodont reptiles are divided into several groups, sub-orders, and families, of which we will take first those known as Dicynodonts,[1] because they illustrate the anomalous nature of the teeth, to which we referred just now.

Sir R. Owen, after a careful study of these interesting remains concluded[2] that there had formerly existed in South Africa—probably in a great lake or inland sea—a race of reptiles presenting, in the construction of their skulls, characters presented by the crocodile, the tortoise, and the lizard, but possessing a pair of long tusks implanted in distinct sockets, like those of the walrus. No other kind of teeth had they; and, as in the case of a tortoise, the lower jaw appears to have been armed by a trenchant sheath of horny matter. The tusks are of a finer texture than that of the crocodile's teeth, and almost as dense as in the hyæna (see Plate X.). These illustrations are taken from photographs very kindly sent by Professor Amalitzky, of Warsaw, who has of late years laboriously worked out nodules containing bones, skulls, etc., from Permian strata on the banks of the Northern Dwina, near Archangel; the results of his researches are of great value (see Sir E. Ray Lankester's *Extinct Animals*, p. 212). The head is shown in Fig. 26, and the fore limb in Figs. 27, 28. As it is, at present, hardly possible to restore the skeleton with any degree of certainty, we have not ventured to show the Dicynodon in our Plate of New Red Sandstone Reptiles (see Plate IX.). A carnivorous genus is shown in Plate XI.

[1] The genus *Dicynodon* is so called from two Greek words: *dis*, twice; and *kunodos*, dog-toothed, on account of the two tusk-like canine teeth in the upper jaw.

[2] Professor Owen's *Memoir on the Dicynodon*, *Geol. Trans.*, second series, vol. vi., with plates.

Some of the arguments used by Owen in his *Memoir*, above referred to, may be briefly condensed as follows: The creatures to which these skulls belonged were not mammals (although to an evolutionist they are a foreshadowing of that class), both on account of the double nasal apertures, one of which is seen in Fig. 26; and also because no mammal has a brain-cavity so relatively small. They were not crocodiles, as indicated by certain other features of the skull. They were not turtles or

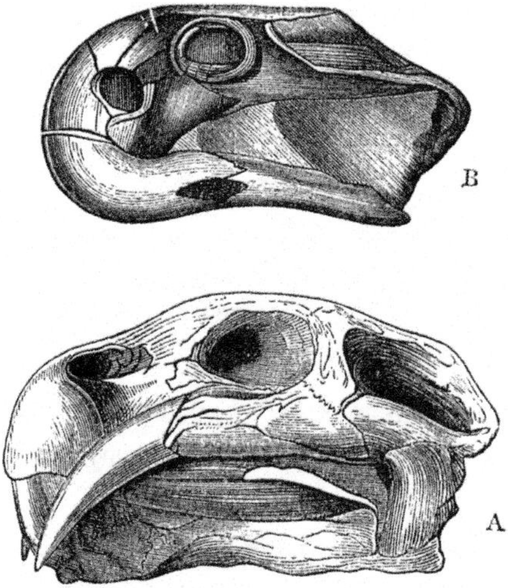

FIG. 26.—Anomodont skulls. A, skull of *Dicynodon*; B, skull of *Oudenodon*, from Karoo strata, South Africa. (After Owen.)

tortoises, for all such reptiles have a single nasal opening placed in the middle of the fore part of the skull. They could not be fishes, for fishes breathe in quite a different way. Neither could they be batrachians (frogs, etc.), nor yet snakes, as is also proved by the structure of their skulls. Certain other features of the skull show a relationship with the Lacertilians, or true lizards. The skulls are mostly of small size; but that of Dicynodon tigriceps is as much as twenty inches in length. That of

1. SKULL OF DICYNODON. PALATAL ASPECT.

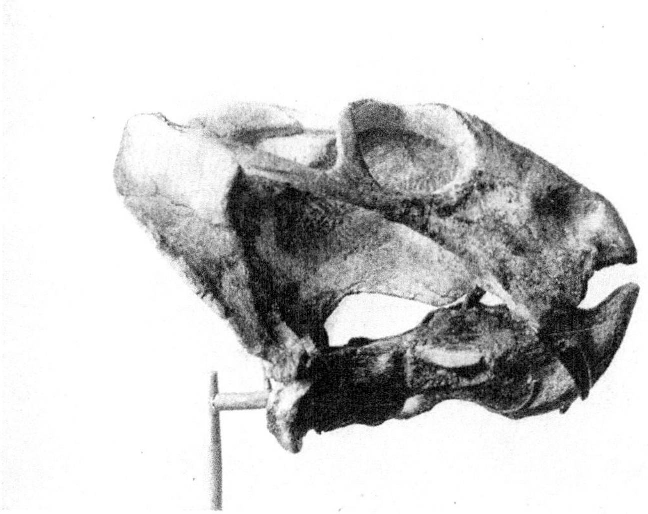

2. SKULL OF DICYNODON TRAUTSCHOELDI.
From photographs kindly sent by Prof. Amalitzky of Warsaw.

PLATE X.

D. lacerticeps ("lizard-headed," Fig. 26) is only six inches long. Dicynodonts have since been found in the Gondwana series in Central India, and in the Elgin Sandstone, Scotland.

The vertebræ, or joints of the backbone, are hollow on both sides (biconcave)—a feature common to fishes and Labyrinthodonts. Probably they were good swimmers, and spent much of their time in the water; but it is quite clear that they

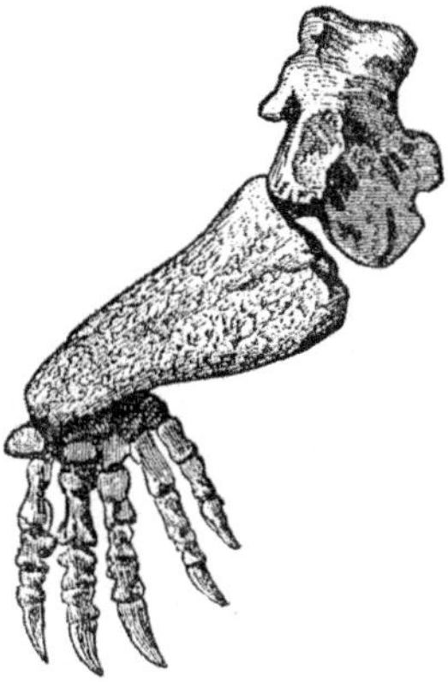

FIG. 27.—Fore limb of *Dicynodon*. (After Owen.)

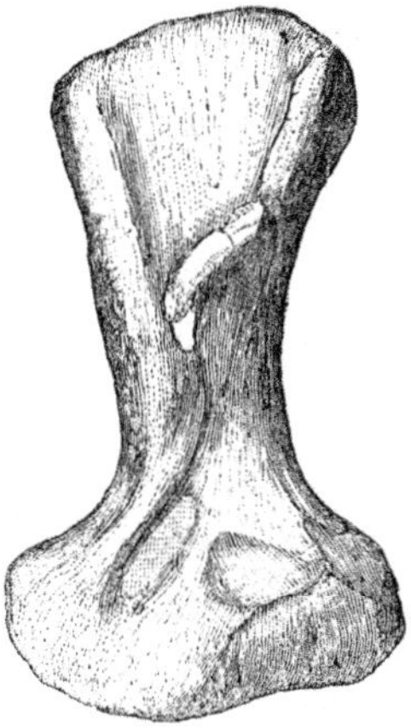

FIG. 28.—Arm-bone (humerus) of *Dicynodon*. ¼ natural size. (After Owen.)

were air-breathers, and so must have come up to the surface to breathe.

The genus Oudenodon,[1] although it has no teeth, so closely resembles the Dicynodon that it must be included in the same family. The skull is shown in Fig. 26. Professor Owen even suggested that the absence of teeth might merely denote a difference of sex; but this view is not accepted by others. The general shape of the skull is very similar to that of Dicynodon, and the

[1] Greek—*ouden*, nothing; *odous*, *odontos*, tooth.

jaws ended in a kind of beak. There are several species, and some attained very large dimensions.

In Oudenodon Bainii the skull is six inches long.

Platypodosaurus[1] is the name given to a considerable portion of the skeleton of a Dicynodont reptile from the Karoo strata of South Africa; but, unfortunately, the skull is unknown, and it may prove to be identical with Oudenodon. The structure of the pelvis, or region of the hips, is remarkably mammalian. The name indicates that it had broad feet.

The genus Endothiodon represents a remarkable family of large reptiles from the same strata, distinguished by the presence of teeth on the palate. The skull resembles that of Oudenodon, but the muzzle is longer.

Two genera, Placodus[2] and Cyamodus, from Triassic strata in Germany, represent another group, the Placodonts, which are very remarkable, but unfortunately at present are only known by their skulls. These, however, are sufficiently peculiar to deserve notice. The teeth represent a curious modification previously unknown in the reptile class, but of which the class of fishes affords numerous examples. In the latter genus the skull is as broad as it is long, and the jaws were very strong. This powerful action of the jaws relates to the form and size of the teeth, which are broad and flat, like paving stones, and evidently adapted to crack and bruise shells of molluscs and crustacea (see Fig. 29). Although now admitted to be a reptile, this remarkable genus was once considered by Agassiz, Owen, and others to be a fish; but where fossil remains are so imperfect, mistakes may easily be made at first.

Tapinocephalus[3] and Titanosuchus[4] are the names given to two

[1] Greek—*platus*, broad; *pous*, *podos*, foot; and *sauros*, lizard.
[2] Greek—*plax*, *plakos*, any flat thing; *odous*, tooth.
[3] Greek—*tapeinos*, low; *cephalos*, head.
[4] Greek—*titan*, a titan; *souchos*, a crocodile.

huge Anomodonts from South Africa, both belonging to the same family, the teeth of which indicate carnivorous habits. An

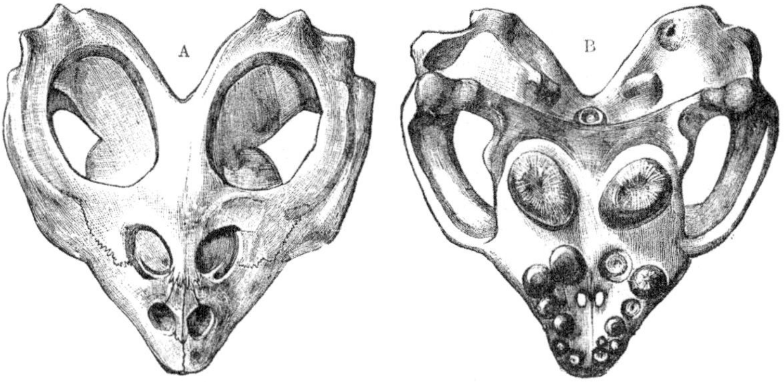

FIG. 29.—Skull of *Cyamodus laticeps*, upper view (A) and palatal view (B), from the Muschelkalk of Baireuth, Germany; ¼ natural size.

imperfect skull, several entire limb-bones, and vertebræ are preserved in the national collection at South Kensington.

Galesaurus,[1] of which the head is shown in Fig. 30, belongs to a remarkable group of Anomodonts, first described by Sir R. Owen,

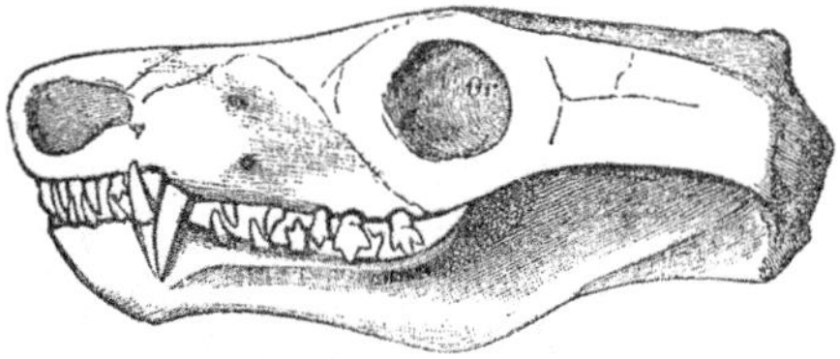

FIG. 30.—Skull and mandible of an Anomodont, *Galesaurus planiceps*. Length 6 inches. Karoo strata, South Africa. (After Owen.)

and called by him Theriodonts,[2] because the form and order of arrangement of their teeth bear a striking resemblance to those

[1] Greek—*galé*, weasel; *sauros*, lizard.

[2] Greek—*therion*, wild beast; and *odous*, *odontos*, tooth: because the teeth resemble those of savage carnivorous creatures, such as lions or wolves.

of carnivorous mammals; for the incisors are separated from the molars by well-developed canines, and the canines of the lower jaw crossed those of the upper in front. In some of the members of this family the upper canine teeth are long and trenchant, and the incisors large and close together. Visitors to the Natural History Museum will find a most valuable set of specimens of the skulls, etc. (Gallery IV.).

In most reptiles, living and extinct, the teeth that are worn away by use, or otherwise lost, are replaced by others that are constantly forming in the jaws; but in the case of Theriodonts there is no evidence of preceding teeth, like the milk-teeth in

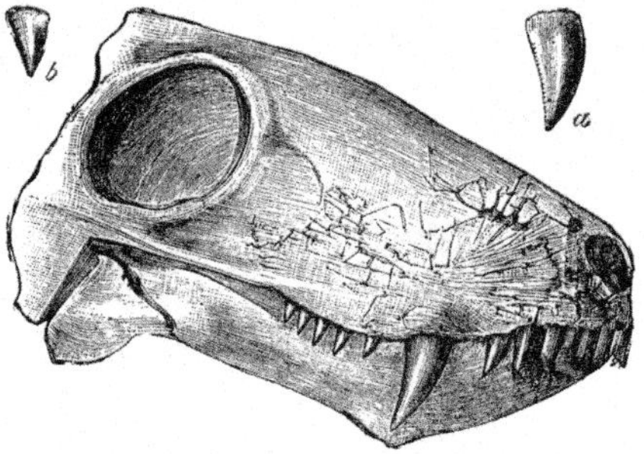

FIG. 31.—Right side-view of skull and mandible of a Theriodont (*Ælurosaurus felinus*), ⅔ natural size, with two upper teeth, natural size (*a*, *b*), from the Triassic Karoo Formation of Beaufort West, Cape Colony. Behind the large orbit the back part of the skull is broken away. (After Owen.)

mammals, nor of succeeding teeth, like the crocodile's. Sir R. Owen therefore concluded that these creatures had but one set of teeth, which lasted through life. He has described eleven genera, varying in the size and form of the skull and teeth; they are all from South Africa, and are figured in his *Catalogue of the Fossil Reptiles of South Africa.*

The work of describing and classifying Anomodont reptiles has

1. SKELETON OF INOSTRANSEVIA ALEXANDRI.

2. SKULL OF INOSTRANSEVIA ALEXANDRI.
From photographs kindly sent by Prof. Amalitzky of Warsaw.

PLATE XI.

since been carried on by the late Professor H. G. Seeley, F.R.S., who had contributed a series of elaborate papers to the Royal Society.[1]

So far we have only spoken of Anomodonts that are imperfectly known; it therefore now remains, before we part company with this wonderful extinct order, to describe one of which the skeleton is practically complete. This is the remarkable Pareiasaurus,[2] for the discovery and description of which palæontologists are

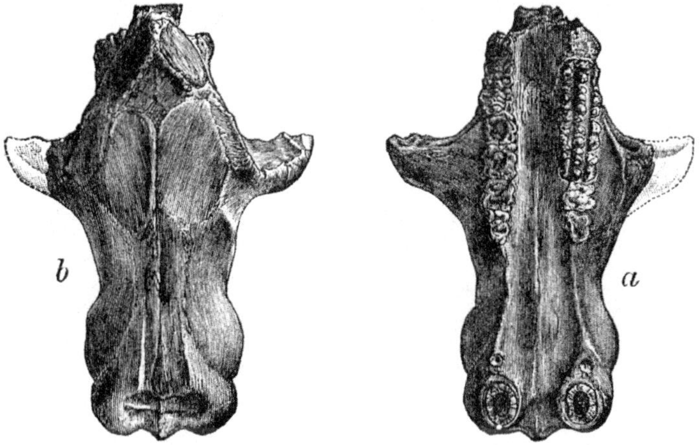

FIG. 32.—Skull of *Tritylodon longævus*, *a* palatal view, *b* upper view. From Triassic strata, Basutoland.

indebted to a distinguished English geologist, the late Professor H. G. Seeley, F.R.S.[3] Visitors to the Natural History Museum will see the unique specimen brought home by Professor Seeley at the end of the fossil reptile gallery, in a glass case by itself (see Fig. 34). The story of its discovery, as given in the journals above

[1] *Philosophical Transactions*, vols. 179, 183, etc.

[2] Greek—*pareia*, the cheek-piece of a helmet; and *sauros*, lizard. So named because the cheek-bones descend so as partly to cover the back of the lower jaw.

[3] *Philosophical Transactions*, vol. 183 B (1892), and *Journal of South African Philosophical Society*, vol. vi. p. 5 (1889–90).

quoted, may be briefly condensed as follows :—In the year 1889 Professor Seeley visited Cape Colony, and examined the Museums of Cape Town and Graham's Town, with a view to studying such remains of the creature as were then known; but almost every specimen of real interest had already been sent to the British

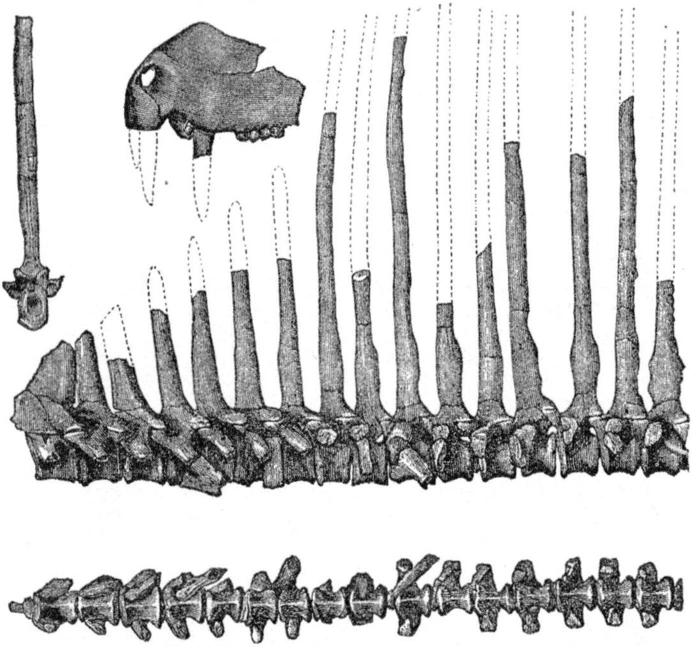

FIG. 33.—Parts of skeleton of *Dimetrodon incisivus*, from Permian strata, Texas. About ¼ natural size. (After Cope.) See p. 118.

Museum. He succeeded in unearthing, in the year 1889, in the Karoo rocks, the most valuable specimen shown in Fig. 34.

There is probably no other South African fossil reptile in which the teeth are developed to the same extent in rows on the palate. Three rows of scutes, or bony scales, appear to have extended down the middle of the back, but the restoration by Professor Amalitzky (Plate XII.) shows many more. Some of these are still preserved in contact with the neural spines on the specimens, but they are not easily detected. Each scute is

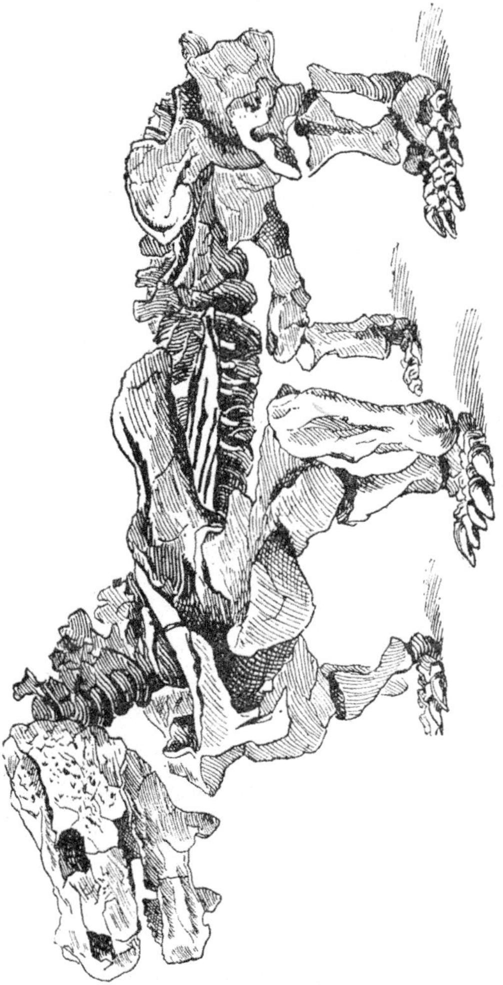

FIG. 34.—Skeleton of a large Anomodont reptile, *Pareiasaurus Baini*, from Karoo strata, South Africa, exhibited in the Natural History Museum. Length 9 feet. (From a sketch by J. R. Hutchinson.)

about two inches wide. There were also scutes on the skull, as in the Labyrinthodonts.

The creature must have been fully nine feet long when alive. The great width and squareness of the animal is a feature that strikes one at once. The skull decidedly resembles that of a Labyrinthodont, and, even to a casual observer, has a frog-like look. But internally it is like that of the Tuatara, or Sphenodon, of New Zealand, which is a survival from very old times, and throws much light on several orders of Triassic reptiles. The backbone, or vertebral column, also resembles that of the little Tuatara, and not that of a Labyrinthodont. In the shoulder girdle we see a resemblance to the Anomodonts, as also in the pelvis, or region of the hips, which is remarkably mammalian. Every one who looks at the skeleton must be struck with the great strength and massiveness of the limb-bones. The arm-bone suggests two very triangular wedges fastened together; so does the femur, or thigh-bone. The present writer suggested three years ago that all this strength may have been for the purpose of supporting a massive shield as in the case of the common tortoise, and now, 1909, he hears that another genus Propappus actually had such an armoured shield. For the sake of comparison, a specimen of the Tuatara (Sphenodon) is placed in a case close by (Wall-case 9). Visitors will find it instructive to compare the skeleton of a Sphenodon with that of the Pareiasaurus. On so doing, they will see a resemblance in the limb-bones of the two creatures. How fortunate that this little lizard should have survived so many geological ages—as it were on purpose to give us some help in studying the long-lost Anomodonts! Professor Seeley considers that the limbs, as represented by their bones, are intermediate between those of mammals and the tailed amphibians, such as newts and salamanders. But the creature was more of a reptile than an amphibian, as is shown by the fact that its skull articulated with the first joint of the backbone by means of one condyle,

1. MODEL OF PAREIASAURUS KARPINSKII.
By Prof. Amalitzky.

2. SKELETON OF PAREIASAURUS KARPINSKII.
From photographs kindly sent by Prof. Amalitzky of Warsaw.

PLATE XII.

or knob-like process, instead of by two, as in the Labyrinthodonts and all the amphibia.

Palæontologists have concluded that Labyrinthodonts, Anomodonts, and Monotreme mammals are all descended from some common amphibian stock. If this theory is true, the Anomodonts retained, while they were yet in existence, structures showing a relationship with mammals or a foreshadowing of them, which in all other reptiles have been altogether lost.

Palæontologists are often greatly helped in their studies by comparing extinct types of life with some of the living types which are very old-fashioned and, one would think, *ought* to have become extinct ages ago, but fortunately have not! We mentioned just now the Tuatara, as a case in point; another still more important case is that of the Monotreme mammals—the curious Duckbill and the Spiny Ant-eater (Echidna) of Australia (which, of course, is not one of the true Ant-eaters). These two remarkable egg-laying mammals show some wonderful points of resemblance to the Anomodonts—points which cannot, however, be explained without entering too much into dry details of Anatomy. Indeed, they are a great puzzle to naturalists, and one hardly knows whether we ought to call them reptiles or mammals; but since they do suckle their young ones (after a primitive fashion, though), they have been placed with the latter; still it requires a slight stretch of imagination to look upon an egg-laying creature as a mammal.

Leaving now the Anomodonts, we pass on to consider another order—the Proterosauria. In America and France the subject has been studied by Professors Cope[1] and Gaudry. The former palæontologist has described, amongst others, such remarkable

[1] *History of the Vertebrata of the Permian Formation of Texas*, by Professor E. D. Cope, Pal. Bull, No. 32.

forms as Eryops (Fig. 20), Empedias, and Dimetrodon. As these are not completely known, we have refrained from attempting to restore them. Perhaps the most remarkable is the Dimetrodon[1] (see Fig. 33) belonging to the order Pelycosauria. Not only was the skull provided with formidable tusks, but the vertebræ of the back present a new feature in having very long neural spines. In one species the height of the spine is actually more than twenty times the length of the centrum! According to Professor Cope, these spines formed a kind of elevated fin on the back, of which it is difficult to imagine the use; but then there are many living animals with bony structures which, if only known in a fossil state, would greatly puzzle every one. Plate XIII. shows a photograph of a restoration of the allied Naosaurus from Permian strata of Texas, for which the writer is indebted to Professor H. F. Osborn. We will take Proterosaurus next.

The skull is more lizard-like than crocodilian. There are mainly two interpretations of this fossil. (1) Cuvier and Huxley class it unreservedly with the lizards. (2) Von Meyer and Owen consider that it represents a new type of reptile. Von Meyer published an elaborate monograph on the subject, with nine folio plates (1856). Professor Seeley also considers that it shows no particular affinity with any living group of reptiles. It evidently came of a very ancient stock. With regard to its habits, Owen concluded that it was an aquatic animal. He says, "The strength of its neck and head, and the sharpness of its teeth, enabled it to seize and overcome the struggles of the active fishes of the waters which deposited the old Thuringian copper-slates." There can be but little doubt, however, that it was capable of walking on land as well as of disporting itself in the water Certain ganoid fishes

[1] Greek—*dis*, double; *metron*, measure; *odous*, tooth.

MODEL OF NAOSAURUS RESTORED. FROM PERMIAN STRATA OF TEXAS.

PLATE XIII. Reproduced by permission of the American Museum of Natural History and Prof. Henry Fairfield Osborn.

abounded in the seas of the Permian period. These, as Owen points out, may have been its prey.

The beak-headed lizards, or Rhynchocephalia,[1] belong to another very remarkable order of reptiles that lived during the New Red Sandstone period. The order is now almost extinct, being only represented by the Tuatara, or Sphenodon, of New Zealand.

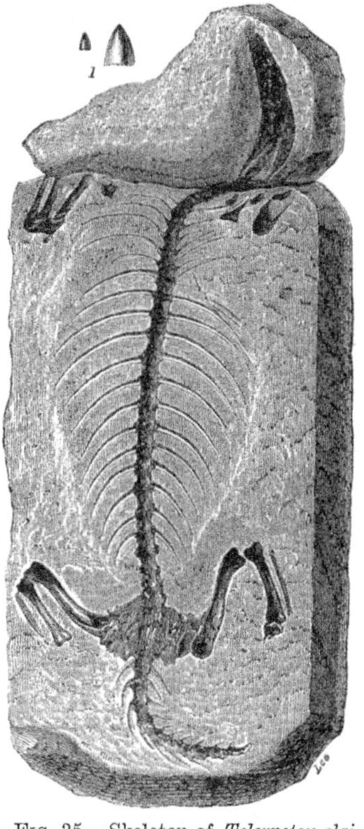

FIG. 35.—Skeleton of *Telerpeton elginense*, from the Elgin Sandstone. Natural size. 1. Tooth. (After Mantell.)

There are three extinct reptiles to be considered under this head before we conclude our survey of Triassic reptiles, and first we will take the little Telerpeton. It was found in the New Red Sandstone strata of Elgin—then considered to be the Old Red Sandstone, and a great controversy took place with regard to their geological age. But now most geologists accept the view that they are of New Red Sandstone age, as asserted by Professor Huxley, and others. The discovery of Telerpeton was an important event in the scientific world at the time, chiefly because the Elgin sandstones were then considered to be of Old Red Sandstone age.

The name now generally accepted denotes the remote antiquity of this reptile.[2]

[1] Greek—*runchos*, beak; *cephalos*, head. [2] Greek—*tele*, far off; *erpeton*, reptile.

A model or cast of this delicate fossil is to be seen in the Natural History Museum. Unfortunately, the head is not preserved. Dr. Mantell considered Telerpeton to be partly amphibian and partly saurian. But, some years later, Professor Huxley came to the rescue, and threw more light on the subject. In a paper read before the Geological Society,[1] he described a new specimen, which showed the skull. This specimen, the property of Mr. James Grant, of Lossiemouth, was sent to him by that veteran geologist, the late Rev. Dr. Gordon, of Birnie, by Elgin, who, for fifty years or more, had taken a deep interest in the geology of the district. It is larger, being ten to twelve inches

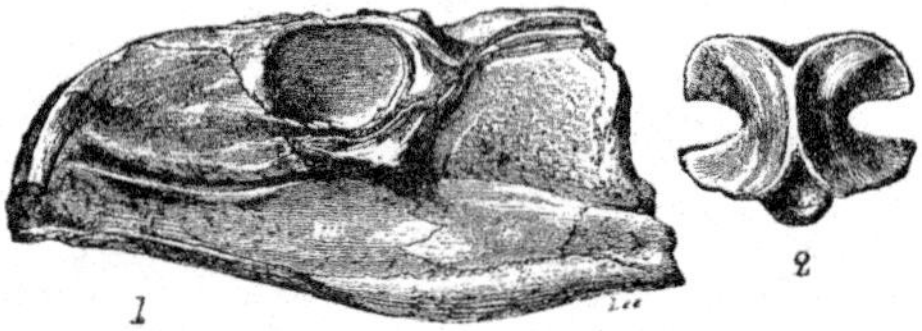

FIG. 36.—Skull of *Rhynchosaurus articeps* from the Trias, from the New Red Sandstone near Shrewsbury; ½ natural size. 2. Upper aspect of a dorsal vertebra; natural size. (After Mantell.)

in length. As is usual with fossils from the Elgin sandstone, the bones are represented by casts, but these were well defined. Professor Huxley was able to show that Telerpeton has no relationship with amphibians. In all its characters it is a true lizard. Instead of being a low, or "generalised" form, it is somewhat "specialised," or highly developed.

The second example of the reptilian order now under consideration is the Rhynchosaurus, discovered by Dr. O. Ward, in a quarry in the New Red Sandstone, at Grinsill, near Shrewsbury. His discovery comprised the skull and a considerable portion of the skeleton (see Fig. 36). Professor Owen also found some remains of the same animal in this quarry.

[1] Published in the *Journal* of 1867, vol. xxiii. p. 77.

The animal was about three feet in length. The general aspect of the skull differs from that of modern lizards, and resembles that of a bird or a turtle, which resemblance is increased by the apparent absence of teeth. There was, however, a single row of teeth on the palate. Professor Owen concluded that this ancient creature had its jaws encased in a bony or a horny sheath. The fortunate preservation of the skull has brought to

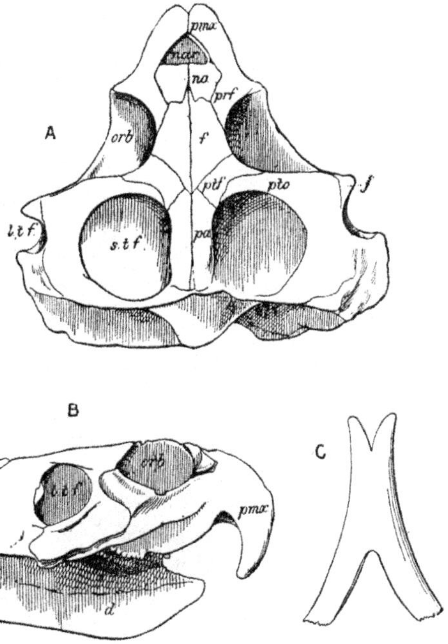

FIG. 37.—Skull and mandible of *Hyperodapedon Gordoni*. A. Upper view; B. Side view; C. Mandible. From the Elgin Sandstone of Triassic age. (After Dr. A. Smith Woodward.)

light a curious modification of the lizard type of structure, that seems to lead on towards birds and turtles.

The Hyperodapedon was a much larger creature. It was a terrestrial reptile, some six or seven feet long, and does not appear to have been armed with scutes or spines of any kind.

A fine specimen is to be seen at the Natural History Museum (Wall-case 9). Like the Rhynchosaurus, it had a broadly triangular form of skull, with the orbits for the eyes directed upwards, and the jaws prolonged into a sharp curved beak, so that it comes into the same order of "beak-headed lizards." But, unlike the latter reptile, it had several rows of well-developed, low, conical teeth, both on the jaws and palate. The specimen at South Kensington shows the head, neck, backbone, and ribs, together

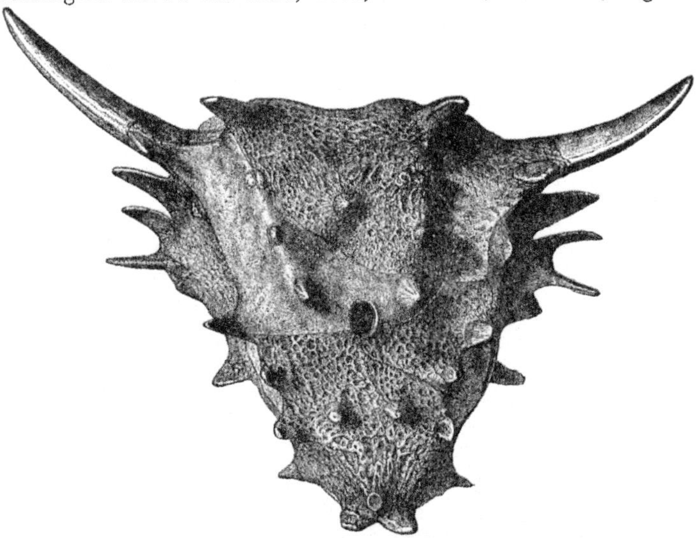

FIG. 38.—Front view of skull of *Elginia mirabilis*, as restored from natural moulds in the Elgin Sandstone, by Mr. E. T. Newton.

with some of the limb-bones. We have therefore ventured on a restoration by Mr. Smit (see Plate IX.), in order to give the reader an idea of its curious head and beaked jaws. (The limbs might be larger.) Hyperodapedon must have had a wide geographical range during the period of the Trias; for its remains have also been found in the centre of England (Warwickshire), in Devonshire, and in Central India. A much larger species from India is reckoned to have attained a length of seventeen feet.

Professor Huxley concluded that Hyperodapedon and Rhynchosaurus do not depart from the ordinary lizard type of structure more than some of our modern lizards do—such as the Monitor, Chameleon and Gecko, to which both are allied. He thinks that, even in Triassic times, the lizard type had become highly specialised.

Elginia mirabilis is the name proposed for the skull of a reptile

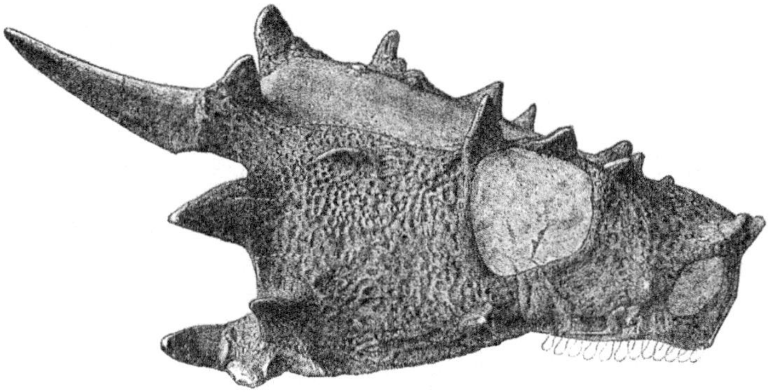

FIG. 39.—Side view of skull of *Elginia mirabilis*, as restored from natural casts in the Elgin Sandstone, by Mr. E. T. Newton.

which, on account of the extreme development of horns and spines, reminds one of the living lizards, Moloch and Phrynosoma. The skull of this ancient saurian (see Figs. 38, 39) is unlike that of any living or fossil form, and seems to show affinities with both the Labyrinthodonts and the lizards. All the remains are in the form of hollow moulds in blocks of stone, but Mr. Newton has, with wonderful skill and patience, made casts in guttapercha and so reconstructed the skull, as shown in our illustrations. The Elginia was closely related to Pareiasaurus.

CHAPTER VIII

THE DRAGONS OF OLD TIME—DINOSAURS

"What we know is but little; what we do not know is immense."—La Place.

Was there ever an age of dragons? Tradition says there was; but there is every reason to believe that the fierce and bloodthirsty creatures, of which such a variety present themselves, are but creations of the imagination,—useful in their way, no doubt, as pointing a moral or adorning a tale, but, nevertheless, wholly without foundation in fact. The dragon figures in the earliest traditions of the human race, and crops up again in full force in European mediæval or even late romance. But in spite of all the manifest absurdities of the dragons of various nations and times, geology reveals to us that there once lived upon this earth reptiles so great and uncouth that we can think of no other but the time-honoured word "dragon" to convey briefly the slightest idea of their monstrous forms and characters.

So there is some truth in dragons, after all. But then we must make this important reservation—viz. that the days of these dragons were long before the human period; they flourished in one of those dim geological ages of which the rocks around us bear ample records.

It is a strange fact that human fancy should have, in some cases at least, created monsters not very unlike some of those antediluvian animals that have, during the present century, been discovered in various parts of Europe and America. Some unreasonable persons will have it that certain monstrous reptiles of the Mesozoic era, about to be described, must have somehow managed to survive into the human period, and so have suggested to early races of men the dragons to which we have alluded. But there is no need for this untenable supposition. By a free blending together of ideas culled from living types of animals it would be very easy to construct no small variety of dragons; and so we may believe this is what the ancients did.

The announcement by Baron Cuvier—the illustrious founder of Palæontology—that there was a period when our planet was inhabited by reptiles of appalling magnitude, with many of the features of modern quadrupeds, was of so novel and startling a character as to require the prestige of even his name to obtain for it any degree of credence. But subsequent discoveries have fully confirmed the truth of his belief, and the "age of reptiles" is no longer considered fabulous. This expression was first used by Dr. G. Mantell as the title of a paper published in the *Edinburgh Philosophical Journal* in 1831, and serves to remind us that reptilian forms of life were once the ruling class among animals.

The Dinosaurs are an extinct order comprising the largest terrestrial and semi-aquatic reptiles that ever lived; and while some of them in a general way resembled crocodiles, others show in the bony structures they have left behind a very remarkable and interesting resemblance to birds of the ostrich tribe. This resemblance shows itself in the pelvis, or bony arch with which the hind limbs are connected in vertebrate or back-boned animals, and in the limbs themselves. This curious fact, first brought into notice by Professor Huxley, has been variously interpreted by anatomists; some concluding, with Professor Huxley, that birds are descended from Dinosaurian ancestors; while others, with Professor Owen, consider the resemblance accidental, and in no way implying relationship. Huxley has proposed the name Ornithoscelida, or bird-legged, for these remarkable reptiles.

Dinosaurs must have formerly inhabited a large part of the primæval world; for their remains are found, not only in Europe, but in Africa, India, North and South America, and even in Australia; and the geologist finds that they reigned supreme on the earth throughout the whole of the great Mesozoic era. Their bodies were, in some cases, defended by a formidable coat of armour, consisting of bony plates and spines, as illustrated by the case of Polacanthus (p. 173), thus giving them a decidedly dragon-like appearance. The vertebræ, or bony segments of the back-bone, generally have their centra hollow on both ends, as in the Ichthyosaurus; but in the neck and tail they are not unfrequently hollow on one side and convex on the other. In some of the largest forms the vertebræ are excavated into hollow chambers. This is apparently for the sake of lightness; for a very large animal with heavy solid bones would find it difficult to move freely. In this way strength was combined with lightness.

All the Dinosaurs had four limbs, and in many cases the hind pair were very large compared to the fore limbs. They varied enormously in size, as well as in appearance. Thus certain of the smaller families were only two feet long and lightly built; while others were truly colossal in size, far out-rivalling our modern rhinoceroses and elephants.

The limbs of Cetiosaurus, for example, or of Stegosaurus, remind us strikingly of those of elephants. The celebrated Von Meyer was so struck with this likeness that he proposed the name Pachypoda for them, which means thick-footed. Professor Owen opposed this name; for it was misleading, and only applied to a few of them. He therefore proposed the name we have already been using, viz. Dinosauria,[1] and this name has been generally retained. We are thus led to connect them with lizards and crocodiles, rather than with birds or quadrupeds. The strange and curiously mixed characters of the old-fashioned reptiles is forcibly illustrated by these differences of opinion among leading naturalists. The late Professor Seeley, another authority, refused to consider them as reptiles, at least in the ordinary sense of the word.

Extinct forms of life are often so very different to the creatures inhabiting the world of to-day, that naturalists find it a hard task to assign them their places in the animal kingdom. The classes, orders, and families under which living forms are grouped are often found inadequate for the purpose, so much so that new orders and new families require to be made for them; and then it is often quite impossible to determine the relations of these new groups to the old ones we are accustomed to. Dinosaurs offer a good example of this difficulty. Were they related to ancient crocodiles? No one can say for certain; but it is quite possible, and even probable. Again, did certain long-legged Dinosaurs eventually give rise by evolution to the running birds, ostriches, emeus, etc.? This, although supported by weighty authority, is a matter of speculation: we ought to be very careful in accepting such conclusions. The following reflections will serve to show that the fact may be explained in another theory. Some Dinosaurs walked, in a more or less erect position, on their hind limbs. Similar habits, if continued for age after age, must tend to produce somewhat similar appearances, at least externally. Take, for example, the case of whales and

[1] Greek—*deinos*, terrible; *sauros*, lizard.

fishes; the original land mammal from which whales are descended has, in the course of time, become so fish-like in appearance that even in these modern days there are some who yet speak of them as fishes! The shape of a whale is fish-like; it has lost its hind limbs through disuse; it has changed its fore limbs into paddles, which have a certain fin-like aspect; and its cousin, the porpoise, has developed a big triangular fin on the back. To take another example: Pterodactyls are reptiles which acquired the power of flight; and some, at least, of them must have borne a resemblance to modern bats. And yet pterodactyls and bats are in no way related. The pterodactyl was a lizard that could fly; while the bat is a mammal that can do the same. There is no question of relationship. Now let us apply the same kind of reasoning to Dinosaurs and birds. For our part, we confess to being not quite convinced. Dinosaurs walked, ran, and perhaps some of the smaller ones even hopped on their hind legs; would it not follow as a matter of course that certain rather bird-like developments would follow? Let us reason the matter out. For a horizontal body, like that of a lizard, to be properly poised on two legs instead of four, the weight of the viscera must be transferred backward, and the forward, or anterior part of the body lightened. The lower bones of the region of the hips (pelvis), with the contained organs, are thrown backward, while the fore part of the body and the fore limbs are lightened and much reduced in proportionate size.

Changes such as these might be supposed capable of producing those bird-like features in Dinosaurs which we have already noticed, and on the strength of which some authorities believe birds and Dinosaurs to have had a common ancestry. It may perhaps be safer to look upon the ancestry of birds as one of those problems that remain to be solved by some future generation with fuller knowledge derived from first evidence. We insert here an illustration (see Fig. 40) in which are shown, side by side, the limb-bones and those of the region of the pelvis in a crocodile, a dinosaur, and a bird. (The middle drawing is not complete and the leg of the crocodile should be more bent.) Professor Marsh has described some remains of certain small and slender little Dinosaurs (Hallopus) about the size of a fowl, which, from the construction of their feet, he believes were

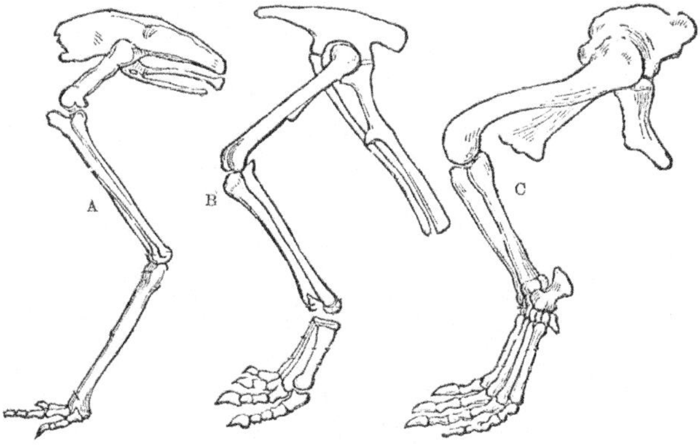

FIG. 40.—Hind limbs of (A) Bird, (B) Dinosaur, and (C) Crocodile.

capable of jumping. It is a question not yet decided whether Dinosaurs laid eggs.

The late Sir Richard Owen, in his great work on *British Fossil Reptiles*, made some valuable remarks on the probable habits of Dinosaurs, which deserve careful attention. He points out that actual observation of a swimming crocodile testifies to the fact that the fore-limbs are laid flat and motionless upon the sides of the chest. Their chief swimming organ is, of course, the tail, which is both long and powerful, and, in some forms, is flattened in a vertical direction. Most of the Dinosaurs at present known have the crocodilian swimming organ; and some, such as the

Claosaurus (p. 169), and the Iguanodon (p. 164), reveal it considerably compressed vertically. In those forms it must have been a cumbrous impediment in the way of walking on land; but in others, such as Anchisaurus (p. 142), the tail was rounder and more lizard-like, and doubtless these carnivorous forms were less hampered in their movements on land. Concerning the smallness of the fore limbs, Professor Owen suggests that the forward dash of the creature through the water in pursuit of fishy prey would be facilitated by this reduction in size of the fore limbs, which would take no part in movement, unless in the way of steering.

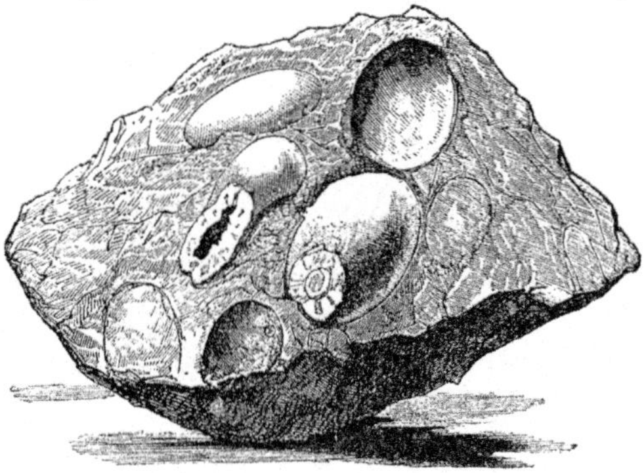

FIG. 41.—Eggs of reptiles from the Great Oolite.

On land they might be of some use in scratching out the nest for the eggs, if they did lay eggs, as well as in helping to support the body.

When any tribe of plants or animals becomes very flourishing, and spreads over the face of the earth, occupying regions far apart from one another, where the geographical and other conditions, such as climate, are unlike, its members will inevitably develop considerable differences among themselves.

During the great Mesozoic period, Dinosaurs spread over a large part of the world; they became very numerous and powerful. Just as the birds and beasts (quadrupeds) of to-day show an almost endless variety, according to the circumstances in which they are placed, so that great and powerful order of reptiles we are now considering ran riot, and gave rise to a variety of forms, or types.

Various attempts have been made to classify Dinosaurs and arrange them in family groups. But considering how imperfect our knowledge is of this very great and important order of reptiles, it is well to bear in mind that all such attempts are provisional and temporary. Much might be said in favour of a complete rearrangement of the classification usually adopted, which is that of the late Professor Marsh. As at present constituted, they appear to make a very artificial group.

The order is usually divided into three sub-orders, as follows: (1) Theropoda, or beast-footed; (2) Sauropoda, or lizard-footed; (3) Predentata, Ornithopoda, or Orthopoda. We shall follow this arrangement (used by von Zittel and others) but merely for convenience. Dr. H. Gadow raises them into a sub-class, making the above groups into orders, which perhaps is better.

But it may be that Dinosaurs are not a homogeneous group. Perhaps some day the sub-order Sauropoda may be taken away altogether and placed with, or at least, near to the crocodiles, which they resemble in some ways. But Dinosaurs are strange beasts, very puzzling to the naturalist; and, much as our knowledge of them has been extended, future discoveries may bring to light forms quite as strange as those already known. Doubtless hundreds of specimens lie awaiting discovery in the unexplored fields of Asia, Africa, and South America. Dinosaurs also show obvious resemblances to the Rhyncocephalia and the Theromorpha. But nobody can say exactly where they should be placed. If we try to construct a genealogical tree of the reptiles we must put the branch called Dinosaurs somewhere near crocodiles, the sphenodon, and even the turtles.

We will take each sub-order in turn, and describe its leading features, and some of the individual genera will be described somewhat fully. In our previous edition the English and the American Dinosaurs were separately dealt with; in the present edition they are not so divided. We have been obliged to omit details here and there to save space, as we are now limited to one volume, and a great part of *Creatures of Other Days*, will be found to be included in the present work. The Theropoda (or beast-footed sub-order) include such forms as Megalosaurus, Anchisaurus, Compsognathus, etc. They played the part of lions and tigers of the present day.

To begin with Megalosaurus. In the year 1824 that keen observer and original thinker, the Rev. Dr. Buckland, described to the Geological Society of London some remains of a very strange and formidable reptile found in the Limestone of Stonesfield, near Woodstock (about twelve miles from Oxford). This rock, known as "Stonesfield slate" from its property of splitting up into thin layers, has long been celebrated for its fossil remains, and from it have been also obtained the bones of some early mammals. It is a member of the Lower Oölitic group.

The portions of skeleton originally discovered consisted of part of a lower jaw, with teeth, a thigh bone (femur), a series of vertebræ of the trunk, a few ribs, and some other fragments. The name Megalosaurus, or "great lizard," suggested itself both to Dr. Buckland and Baron Cuvier, because it was evident from the size of the bones that the creature must have been very big. It is true these bones were not found together in one spot; but Professor Owen came to the conclusion that they all belonged to the same species.

Plate XIV. A LARGE-HORNED DINOSAUR, CERATOSAURUS, FROM NORTH AMERICA. JURASSIC PERIOD.

No entire skeleton of the Megalosaur has ever been found, but there was enough material to enable Dr. Buckland, Professor Owen, and Professor Phillips to form a very fair idea of its general structure. But Ceratosaurus (p. 137) is completely known. It should be mentioned here that Dr. Mantell, the enthusiastic geologist to whose labours palæontologists are greatly indebted, had previously discovered similar teeth and bones in the Wealden strata of Tilgate Forest. Sherborne, in Dorset, is another locality which has yielded a fine specimen of parts of both jaws with teeth. A cast of this may be seen in the geological collection at South Kensington.

The portion of a lower jaw in the Oxford Museum is twelve inches long, with a row of nine teeth, or sockets for teeth. The structure of the teeth leaves no doubt as to the carnivorous habits of the creature. With a length of perhaps thirty feet (see p. 139), capable of free and rapid movement on land, with strong hind limbs, short head, with long pointed teeth, and formidable claws to its feet, the Megalosaur must have been without a rival among the carnivorous reptiles on this side of the world. It probably walked for the most part on its hind legs, as depicted in our illustration, and Professors Huxley and Owen, on examining the bones in the Oxford Museum, were much impressed with the bird-like character of some parts of the skeleton, showing an approach to the ostrich type. The form of the teeth, as pointed out by Dr. Buckland, exhibits a remarkable combination of contrivances.

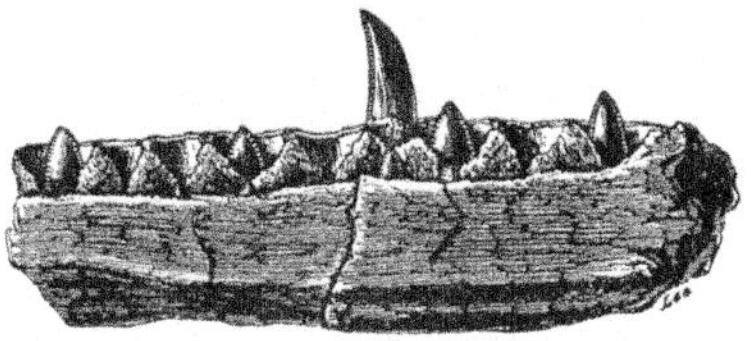

FIG. 42.—Lower jaw-bone of *Megalosaurus*, with teeth.

The genus Megalosaurus—now rendered classic through the

labours of Professors Buckland, Phillips, Owen, and Marsh—may be regarded as the type of the carnivorous Dinosaurs; and it affords an excellent and instructive instance of the gradual restoration of the skeleton of a new monster from more or less fragmentary remains. Certain very excusable errors were at first made in the restoration, but these have since been rectified by a comparison with the allied American forms, such as Allosaurus (see Plate XV.), of which nearly entire skeletons have of late been discovered in strata of Jurassic age—in fact, the same rock in

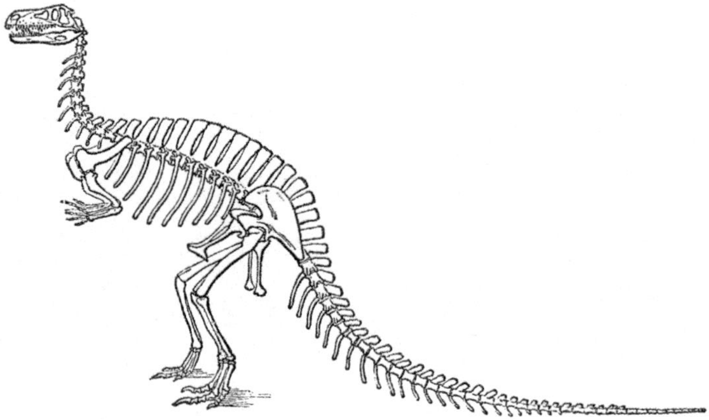

Fig. 43.—Skeleton of *Megalosaurus*, restored from the English Oolites.

Colorado as that in which the huge Atlantosaurus bones lay hid. The accompanying woodcut (Fig. 43) shows how the skeleton has been restored in the light of these later discoveries of Professor Marsh. The large bones of the limbs were hollow, and many of the vertebræ, as well as some of those of the feet, contained cavities, or were otherwise lightened in order to give a greater power of rapid movement.

It is not very difficult to imagine a Megalosaur lying in wait for his prey (perhaps a slender, harmless little mammal of the

SKELETON OF HUGE CARNIVOROUS DINOSAUR, ALLOSAURUS.

PLATE XV. Reproduced by permission of the American Museum of Natural History and Prof. Henry Fairfield Osborn.

ant-eater type) with his hind limbs bent under his body, so as to bring the heels to the ground, and then with one terrific bound

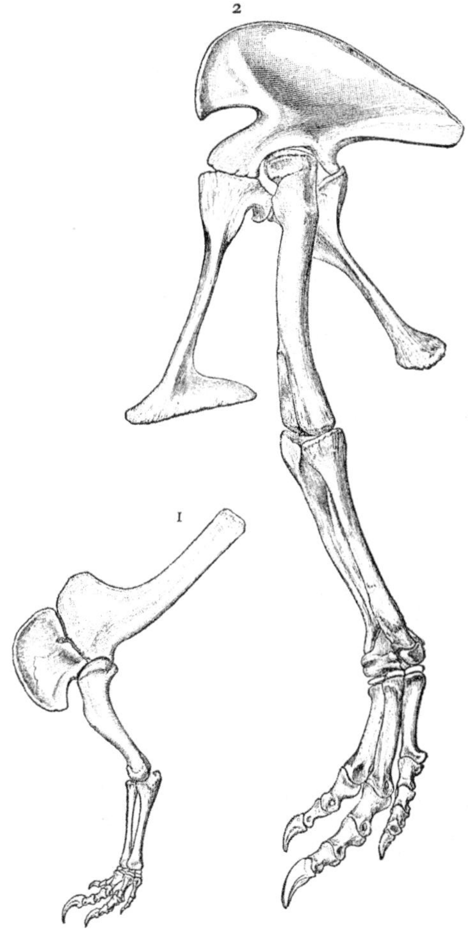

FIG. 44.—Limb-bones of *Allosaurus*. (After Marsh.)
1. Fore leg. 2. Hind leg.

from those long legs springing on to the prey, and holding the mammal tight in its clawed fore limbs, as a cat might hold a mouse. Then the sabre-like teeth would be brought into action by the powerful jaws, and soon the flesh and bones of the victim would be gone!

As we remarked before, the carnivorous Dinosaurs were the lions and tigers of the Mesozoic era, and, what with small mammals and numerous reptiles of those days, it would seem that they were not limited in their choice of diet.

During the early part of the Mesozoic era, at the period known as the Triassic (New Red Sandstone), Dinosaurs flourished vigorously in America, developing a great variety of forms and sizes. Although but few of their bones have as yet been discovered in those rocks, they have left behind unmistakable evidence of their presence in the well-known footprints and other impressions upon the shores of the waters which they frequented (see p. 48).

The late Professor Marsh had the good fortune to discover a very peculiar new form of carnivorous Dinosaur, to which he gave the name Ceratosaurus,[1] because its skull supported a horn. But the horn is not the only feature presented by this interesting creature. Its vertebræ are of a strange and unexpected type; and in the pelvis all the bones are fused together, as in modern birds. Externally, also, the Ceratosaurus differed from other members of the carnivorous group, for its body was partly protected by long plates in the skin, such as crocodiles have: these extended from the back of the head, along the neck, and over the back. An almost complete skeleton was found which indicates an animal about twenty-two feet long. When alive it was probably about half the bulk of the Allosaurus mentioned above. (See Fig. 44.)

[1] Greek—*keras*, horn; *sauros*, lizard. Some authorities consider it to be identical with Megalosaurus.

Seen from above, its skull resembles in general outline that of a crocodile, the facial portion being elongated and gradually

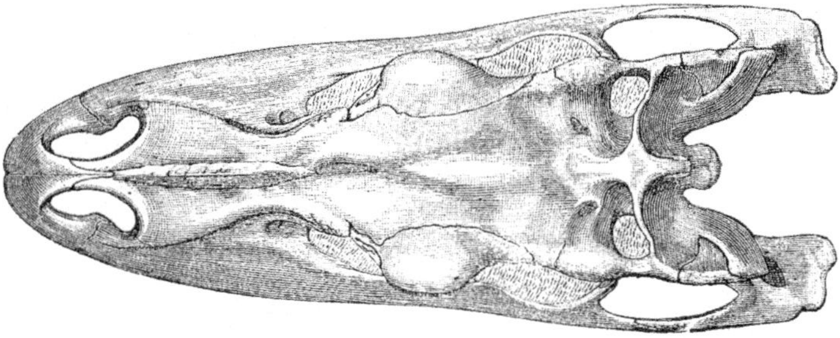

FIG. 45.—Skull of *Ceratosaurus nasicornis*. Top view. (After Marsh.)

tapering to the muzzle, with the nasal openings separate, and placed near the end of the snout.

The teeth of this horned Dinosaur resemble those of the Megalosaur. Its eyes were protected by protuberances of the

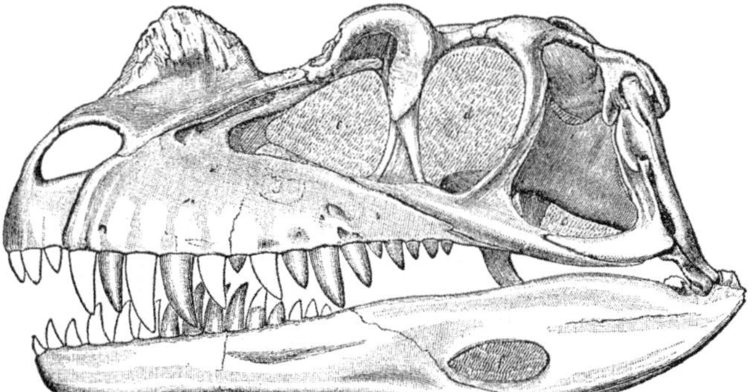

FIG. 46.—Skull of *Ceratosaurus nasicornis*. (After Marsh.)

skull just above the cavity in which the eye was placed (see Figs. 45 and 46). The brain was a good deal larger in proportion to the size of the animal than in Brontosaurus and its allies ; so perhaps we may infer that it was endowed with greater intelligence, as it certainly was more active in its habits. The fore limbs, as in Megalosaurus, were small, and some of the fingers ended in powerful claws, which no doubt it could use to good

FIG. 47.—Skeleton of *Compsognathus longipes*, length about 2 feet. From the Solenhofen limestone.

purpose. This skeleton so much resembles that of Megalosaurus that Mr. Lydekker considers they should be both included in the genus Megalosaurus.

Perhaps the most remarkable of all the Dinosaurs was a

diminutive creature only two feet in length, which was related to those we have just been considering, and whose skeleton has been found almost entire in the now famous Lithographic Stone of Solenhofen in Bavaria. Of this unique type, the Compsognathus, the skeleton of which is in many ways so bird-like, Professor Huxley remarks, "It is impossible to look at the conformation of this strange reptile and to doubt that it hopped, or walked, in an erect or semi-erect position, after the manner of a bird, to which its long neck, slight head, and small anterior limbs must have given it an extraordinary resemblance." (See Fig. 47.)

But by far the greatest of all the carnivorous Dinosaurs was

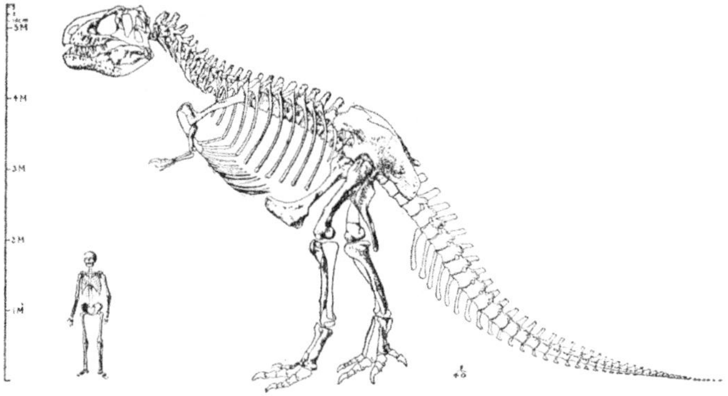

FIG. 48.—Restored skeleton of the huge carnivorous Dinosaur, *Tyrannosaurus rex*. (After Professor H. F. Osborn.) The shaded portions are the actual bones discovered. A scale of meters is given at the side.

the huge and fierce beast known now as Tyrannosaurus rex, a true example of a "terrible lizard," as the word Dinosaur denotes. Its length was about thirty-nine feet, and height about seventeen feet! Plate XV. is a photograph from a restored skeleton of Allosaurus and Fig. 48 gives a diagrammatic view of Tyrannosaurus, with a human skeleton for comparison, and shows which parts were actually discovered, and which are restored, the latter being only in outline. It was discovered in 1902 in Montana, U.S., by the members of an expedition for the American Museum of Natural History in Cretaceous rocks below the Laramie beds. Previous discoveries were made in 1856, 1866, 1868, 1876, 1892, and 1900. A specimen found in the last-named year was described under the name of Dynamosaurus, thus indicating the strength of the beast. It seems to have been contemporary with the Ceratosauria, to be described later on (p. 184). Allosaurus and Laelaps were related forms. The limb-bones in all these forms are hollow.

Another remarkably bird-like form of Dinosaur, also found by Professor Osborn at Como Bluffs, is the Ornitholestes, an active little creature, which seems to have walked on its hind legs only, like a bird. The hind limb is *extremely* bird-like and resembles that of the oldest known bird, Archæopteryx (see p. 213) with its back claw. The hand also is very bird-like, the digits being armed with strong claws. Its bones are hollow (in fact mere shells), another bird-like feature. The present writer was so much struck by the resemblance between this skeleton and that of Archæopteryx, that he even ventured to suggest that possibly this supposed Dinosaur may have been a bird, perhaps the first bird that ever existed!

It is perhaps hardly necessary to remark that to any one thoroughly acquainted with the structures of living animals, a tooth, or a series of teeth, will furnish material from which important conclusions with regard to the structure and habits of an extinct animal may be drawn. So also with regard to some other parts, such as limb-bones, but more especially the bones of which the back-bone is composed (known as vertebræ). Professor Owen has said, "If I were restricted to a single specimen on which to deduce the nature of an extinct animal, I should choose

MODELS OF LAELAPS RESTORED.

Reproduced by permission of the American Museum of Natural History, and Prof. Henry Fairfield Osborn.

PLATE XVI.

a vertebra to work out a reptile, and a tooth in the case of a mammal. Seven or eight different 'characters,'" he says, "may be deduced from a reptilian vertebra." It is of course impossible for any one to reconstruct an entire animal from a single bone, or a few teeth. Not even Owen could do this—in spite of the rather frequent assertions to that effect one sees in newspapers and magazines!

Dr. G. A. Mantell says, "Fossils have been eloquently and appropriately termed *Medals of Creation*," and the eloquent passage by which those words are followed may be transcribed here. He goes on to say, "For as an accomplished numismatist, even when the inscription of an ancient and unknown coin is illegible, can from the half-obliterated effigy, and from the style of art, determine with precision the people by whom, and the period when, it was struck: in like manner the geologist can decipher these natural memorials, interpret the hieroglyphics with which they are inscribed, and from apparently the most insignificant relics trace the history of beings of whom no other records are extant, and ascertain the forms and habits of unknown types of organisation whose races are swept from the face of the earth, ere the creation of man, and the creatures which are his contemporaries. Well might the illustrious Bergmann exclaim, '*Sunt instar nummorum memoralium quæ de præteritis globi nostri fatis testantur, ubi omnia silent monumenta historica.*'"

Geology owes a deep debt of gratitude to the late Dr. Gideon A. Mantell, who, during the intervals of a laborious professional life, collected and described the remains of several strange extinct reptiles, and wrote a number of works on geology, such as served in his day to advance the science to which he was so enthusiastically devoted (see p. 159).

One of the oldest of the Dinosaurs seems to be the

Anchisaurus[1] of Marsh, of which he has made a restoration of the skeleton, as shown in Fig. 49. But it was only by putting together the results of several discoveries that he was enabled to give the complete outline of its skeleton as represented in the accompanying figure, so it may be well to give a brief account of the history of the discoveries that led up to this much-desired result.

As far back as the year 1818, a portion of a skeleton was dis-

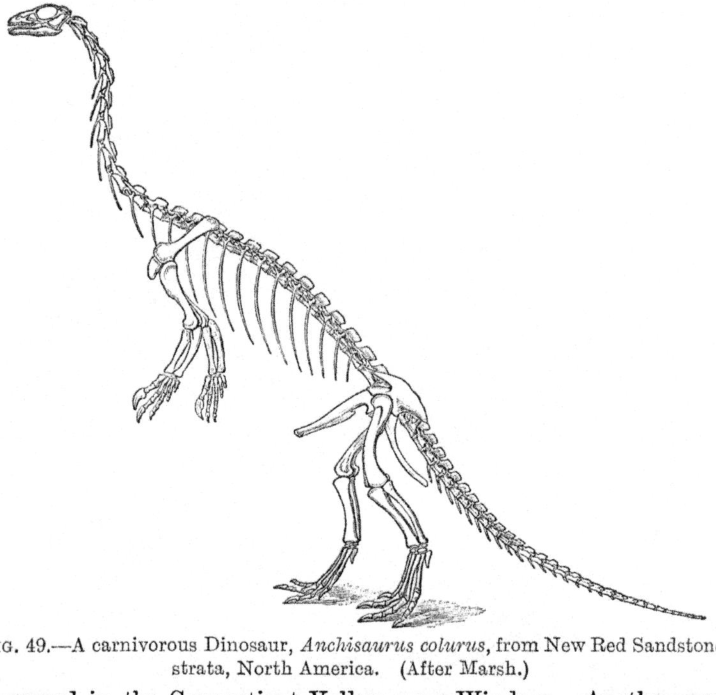

FIG. 49.—A carnivorous Dinosaur, *Anchisaurus colurus*, from New Red Sandstone strata, North America. (After Marsh.)

covered in the Connecticut Valley, near Windsor. Another was found near Springfield, and described by Hitchcock, in 1865, under the name Megadactylus. Later on, in the year 1884, Professor Marsh announced another discovery, near Manchester Connecticut, in almost the same geological level, or "horizon,"—

[1] Greek—*anchi*, near; *sauros*, lizard.

that is to say, in strata of about the same age,—of the remains of an animal of larger size, but in many respects nearly allied to the one described by Hitchcock. Both apparently belong to the same genus. These remains seem to represent an animal about six to eight feet long; but, unfortunately, they are not complete, although when first discovered the skeleton probably was complete, and, with proper care, might have been preserved entire. Another discovery was reported by Professor Marsh, in 1891, when two interesting specimens were secured for the Yale College Museum; one of these was a species of the Anchisaurus. Its skull and the greater portion of the skeleton were found in place. The skull is of moderate size and delicate structure, and in general shape somewhat resembles that of the curious New Zealand reptile known as the Sphenodon, or Tuatara. The vertebræ and limb-bones are hollow, and the whole skeleton was lightly built; the neck and tail were of moderate length.

Writing in the *American Journal of Science* for February of 1893, Professor Marsh reported that the skeletons of five small Dinosaurs had since been discovered in the Connecticut Sandstone. They are sufficiently well preserved to give us most valuable information with regard to all the chief characteristics of the animals to which they once belonged.

With the more complete materials now in his possession, Professor Marsh was enabled to make the restoration of the skeleton of the genus Anchisaurus, seen in Fig. 49. We may attribute a length of about six feet to the creature when alive. The skeleton, from which this restoration was chiefly made, was discovered entire, and apparently in the position in which the animal died. Unfortunately, some of the vertebræ belonging to the neck and tail were lost before the importance of the specimen was realised, but the skull and nearly all the rest of the skeleton was saved. In order to complete the outline, Professor Marsh made use of the corresponding parts, fortunately preserved, in another specimen of an allied species, viz. Anchisaurus solus, found in the same locality. Although it would doubtless be preferable to make restorations from a single specimen only, because the result would be free from all doubt (except perhaps in some of the positions assigned to limbs, etc.), yet this is seldom possible. Those who make extinct forms of life their study are frequently obliged to adopt the method of scholars in reconstructing the text of some classical author, who, when breaks, or *lacunæ*, occur in one manuscript, consult another one which they consider trustworthy, and from it fill in the gaps.

The restoration of this Dinosaur shown in Plate XVII. is based on the figure of the skeleton shown above. The Anchisaurus colurus was one of the most slender and delicate of all the Dinosaurs yet discovered, being only surpassed in this respect by the little bird-like form, from the Solenhofen slate, known as Compsognathus, the skeleton of which is reproduced for comparison on p. 138. This was the creature of which Professor Huxley remarked that no one could look at it without concluding that it must have hopped about on its hind feet, like a bird. The position chosen for the Anchisaurus, as represented in our Plate, is one which, doubtless, the creature was in the habit of assuming during life. But, at the same time, there can be but little doubt that it was also in the habit of progressing on all-fours. In fact, some persons may think that it looks a little "top-heavy," so to speak, in this semi-erect position assigned to it. But, as other carnivorous Dinosaurs of a later period must have walked erect on their hind legs, we may be allowed to follow Professor Marsh in this matter, and to adopt his conclusion.

THE OLDEST-KNOWN DINOSAUR, ANCHISAURUS. FROM NORTH AMERICA. NEW RED SANDSTONE PERIOD.

PLATE XVII.

But the most interesting opinion connected with this discovery is that the Anchisaurus was probably the creature that made some of the tracks found in the strata in which its remains were buried up (see chap. II., p. 44). For half a century or more these tracks have been a fruitful source of contention, and now at last we seem to be literally "on the track" of one of the many antediluvian creatures that made them. This is very satisfactory, and shows that time solves most things. Professor Marsh points out that, on a firm but moist beach, only three-toed impressions would have been left by the hind feet, and the tail could have been kept free from the ground. On a soft muddy shore, the claw of the first toe of the hind foot would have left its mark, and perhaps the tail also might have touched the ground. As we have already remarked, such additional impressions have been found on the Connecticut Sandstones (see Plate I.).

One other carnivorous Dinosaur, named Ammosaurus by Professor Marsh, and allied to the one above described, is also partly known in the same strata, but we cannot at present give any account of it. One or two other forms from Triassic strata are only imperfectly known; such as the Thecodontosaurus, from the Triassic conglomerate of Clifton, Bristol, and Zanclodon, from Germany and South Africa.

Taking one consideration with another, there can be but little doubt that Dinosaurs flourished vigorously during the period of the New Red Sandstone. We have two kinds of evidence to confirm such a conclusion; one is the large number and variety of footprints which they have left behind—although, of course, we cannot suppose that they were all made by Dinosaurs; for, as we have seen, some were probably due to tortoises, while others were probably made by Labyrinthodonts and perhaps even Anomodonts.

Another piece of evidence is the known fact that Dinosaurs were abundant in the succeeding Jurassic period. Judging from the size of some of the footprints, it would appear that not a few of the Triassic forms attained large dimensions. And, if the strata had been more favourable, we should have had more of them preserved. Belodon, described in the author's *Creatures of Other Days* as a crocodile, has been proved by recent discoveries in America to be a Dinosaur of Triassic age.

We pass on to consider the very interesting and huge forms included by the late Professor Marsh (of Yale College), in his suborder Sauropoda, or lizard-footed Dinosaurs. Various parts of the skeletons, such as vertebræ, leg-bones, etc., of these cumbrous beasts have long been known in this country; but Professor Marsh was the first person to discover a complete skeleton.

We shall, therefore, now turn our attention to the bony framework of the huge Brontosaurus (Fig. 50), a vegetable-feeding lizard. But it will be necessary to completely lay aside all our previous notions taken from lizards of the present day, with their short legs and snake-like scaly bodies, before we can come to any fair conclusion with regard to this monstrous beast.

It was nearly sixty feet long, and probably when alive weighed more than thirty-eight tons! that it was a stupid, slow-moving reptile, may be inferred from its very small brain and slender spinal cord. By taking casts of the brain-cavities in the skulls of extinct animals, anatomists can obtain a very good idea of the nature and capacity of their brains; and in this way important evidence is obtained, and such as helps to throw light upon their habits and general intelligence. No bony plates or spines have been discovered with the remains of this monster; so that we are driven to conclude that it was wholly without armour: and, moreover, there seem to be no signs of offensive weapons of any kind except the lash of the tail.

Professor Marsh concludes that it was more or less amphibious

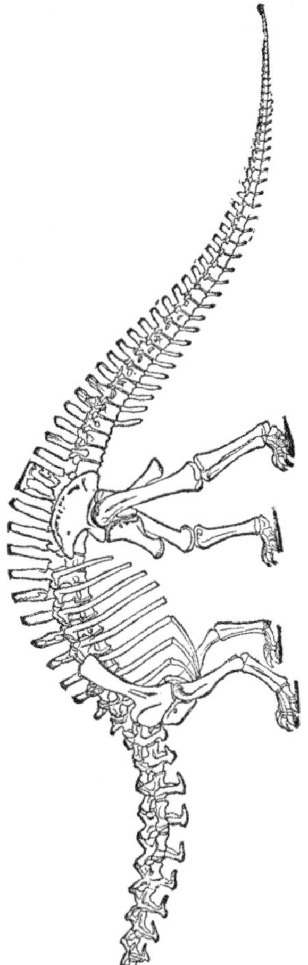

FIG. 50.—Skeleton of a Sauropodous Dinosaur (*Brontosaurus excelsus*), from the Upper Jurassic of Wyoming, U.S.A. (After O. C. Marsh.)

in its habits, and that it fed upon aquatic plants and other succulent vegetation. Its remains, he says, are generally found in localities where the animal had evidently become mired, just as cattle at the present day sometimes become hopelessly fixed in a swampy place on the margin of a lake or river (see p. 19). Each track made by the creature in walking occupied one square yard in extent!

The Sauropoda, or lizard-footed Dinosaurs, show in many ways a decided approach to a simple or generalised crocodile; so much so, that Professor Cope is inclined to include crocodiles and sauropodous Dinosaurs in the same order. Vertebræ, limb-bones, skulls, and teeth have all been discovered through the zeal and energy of Professor Marsh and his comrades, in the far west of America, as well as by the researches of English geologists, assisted by the labours of many ardent collectors of fossils, in this country. Some of these may now be briefly considered.

In Plate XVIII. we have endeavoured to give some idea of a huge thigh-bone (femur) belonging to the truly gigantic Dinosaur called Atlantosaurus. It is six feet two inches long, and a cast of it may be seen in the fossil reptile gallery of the British Museum of Natural History. It should be mentioned, however, that the original specimen is partly restored, so that its exact length to an inch or so is not quite certain. In our illustration it is shown to be a little taller, when placed upright, than a full-grown man. Professor Marsh, the fortunate discoverer of this wonderful bone, calculates that the Atlantosaurus must have attained a length of over eighty feet! and, assuming that it walked upon its hind feet, a height of thirty feet!

It doubtless fed upon the luxuriant foliage of the sub-tropical forests, portions of which are preserved with its remains. Besides this thigh-bone, Professor Marsh has procured specimens of vertebræ from the different parts of the vertebral column; but no skull or teeth. The vertebræ are hollowed out much in the same

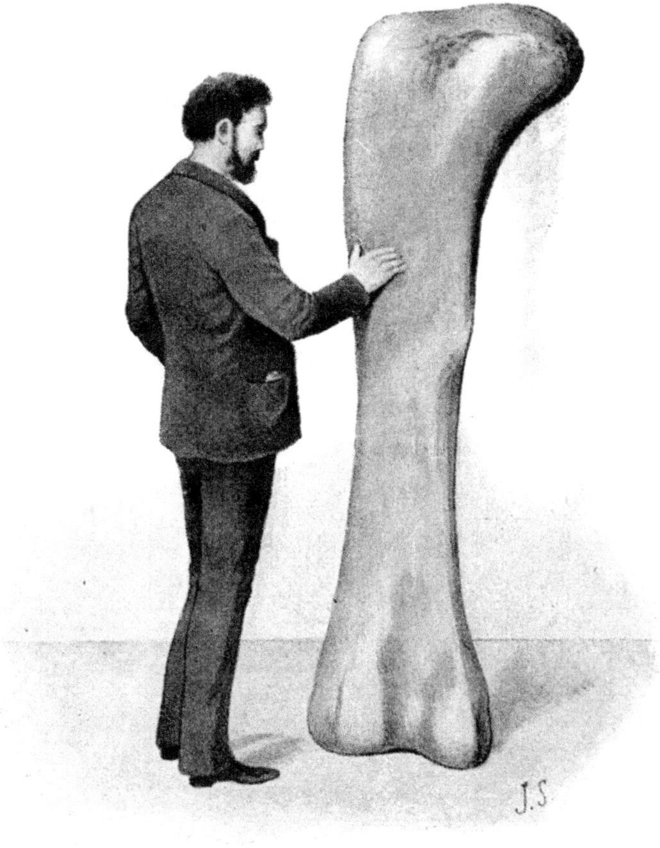

THIGH-BONE OF A HUGE DINOSAUR, ATLANTOSAURUS.

From a cast in the Natural History Museum, London. Length 6 feet 2 inches.

PLATE XVIII.

way as those of Brontosaurus. The fore limbs were large, as in the latter animal; and the extremities of the limbs were provided with claws. It had two sternal bones like Diplodocus (see Plate XXI.). Taking all present evidence, it appears that the Atlantosaurus bore a general resemblance to its smaller contemporary. We can therefore form a fairly good idea of its aspect and proportions.

The same Jurassic strata from the Rocky Mountains have

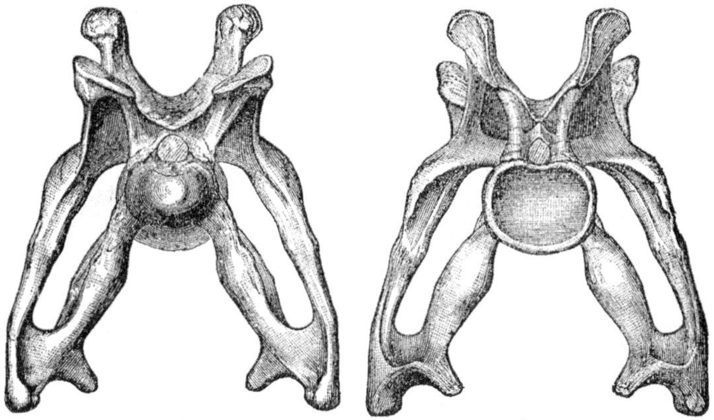

1. 2.

FIG. 51.—Neck vertebræ of *Brontosaurus*.
1. Front view. 2. Back view.

yielded remains of another big Dinosaur, belonging to the same family. This genus, which has been named the Apatosaurus, is represented by a nearly complete skeleton, in the Yale College Museum; and it is fortunately in an excellent state of preservation. Another species, of smaller size, though not so complete, adorns the same collection. This was about thirty feet long, and is known as Apatosaurus grandis.

The remarkably small head is one of the most striking features of these Dinosaurs, and presents a curious contrast to the large and formidable skulls possessed by some other forms to be described further on. But it is clear that no animal with such a long neck as this creature had could have borne the weight of a heavy skull. Short thick necks and heavy skulls always go together. Indeed, the weight of the long neck itself would have been serious had it not been for the fact that the vertebræ in this part of the skeleton, and as far as the region of the tail, have large cavities in the sides of the centra. This cavernous structure of the vertebræ gradually decreases towards the tail. The cavities communicated with a series of internal cavities which give a kind of honeycombed structure to the whole vertebræ. This arrangement affords a combination of strength and lightness in the massive supports required for the huge ribs, limbs, and muscles, such as could not have been provided by any other plan. (See Fig. 51.)

The body of the Brontosaur was comparatively short, with a fairly large paunch (see restoration, Plate XIX.). The legs and feet were strong and massive, and the limb-bones solid. As if partly in order to balance the neck, we find a long and powerful tail, in which the vertebræ are nearly all solid. In most Dinosaurs the fore limbs are small compared to the hind limbs—*e.g.* Megalosaurus, Iguanodon, and Scelidosaurus,—but here we find them unusually large. In this case, then, it is hardly possible that the creature walked upon its hind legs, as many of the Dinosaurs did. There can be little doubt but that many other fierce and formidable Dinosaurs were living at the same time and in the same region with Brontosaurus, whose remains are found in the Jurassic rocks of Colorado (Atlantosaurus beds).

How this apparently helpless and awkward animal escaped in the struggle for existence it is not easy to conjecture; but since there is reason to believe it was more or less at home in the

Plate XIX.

A GIGANTIC DINOSAUR, BRONTOSAURUS EXCELSUS.
Length nearly 60 feet.

water, and could use its powerful tail in swimming, we may perhaps find a way out of the difficulty by supposing that, when alarmed by dangerous flesh-eating foes, it took to the water, and found discretion to be the better part of valour. Although apparently stupid, the Brontosaur probably possessed a good deal of cunning, and we can fancy it stretching its long neck above reeds, ferns, and cycads to get a view of the approaching enemy.

The remains of this interesting Dinosaur (Brontosaurus), which in several ways differs from other members of the "lizard-footed" group, were found in Upper Jurassic beds, near Cãnon City, Colorado. A second smaller species was also discovered near Morrison, Colorado. All the remains lay in the Atlantosaurus beds. These strata—the tomb in which Nature has buried up so many of her dragons of old time—can be traced for several hundred miles on the flanks of the Rocky Mountains, and are always to be known by the bones they contain. They lie above the Triassic strata and just below the Sandstone of the Dakota group. Some have regarded them as of Cretaceous age; but, judging from their fossils, there can be but little doubt that they were deposited during the Jurassic period—probably in an old estuary. They consist of shale and sandstone.

Besides the numerous Dinosaurs, Professor Marsh's colleagues have found abundant remains of crocodiles, tortoises, and fishes, with one Pterodactyl, a flying reptile (see Chap. XI.), and several small marsupials. The wonderful collection of American Jurassic Dinosaurs in the Museum of Yale College includes the remains of several hundred individuals, many of them in excellent preservation, and has afforded to Professor Marsh the material for his classification already alluded to.

For a good many years past the State of Wyoming has been

famous as a happy hunting ground for geologists in search of prehistoric monsters. It was here that Marsh and others found the bones of creatures whose existence had never been suspected, some of which we have already described. But it was in the year 1897 that Mr. Walter Granger, of the American Museum Expedition, discovered the greatest and most wonderful site for fossil bones ever known in the whole history of Science. It simply teems with bones, Dinosaurs, crocodiles, turtles, etc.! Professor H. F. Osborn has been working on this site for six years (from 1898), and has taken therefrom bones weighing nearly 100,000 lbs. (over 44 tons).[1] Judging from the number of thigh bones, these probably represent 73 animals of the following kinds: Giant herbivorous Dinosaurs, 44; plated herbivorous ditto, 3; large carnivorous ditto, 6; small carnivorous ditto, 3; crocodiles, 4; turtles, 5.

But the site is not yet exhausted; it lies not far from Medicine Bow River, and is marked by a ruined hut which some shepherd had built up with blocks of stone containing petrified bones. The illustration on Plate XX. shows a hind limb of a Diplodocus nearly six feet long! from a photograph kindly lent by Professor Osborn. The site is now known as "Bone-cabin Quarry"—a building probably unique in the world's history. At the Como Bluffs (Cliffs), about ten miles off, single animals lie from twenty to one hundred feet apart; but whole skeletons are seldom found. The late Professor Marsh, however, got a nearly complete Brontosaurus here, which is now in the Yale Museum. A geologist is more likely to find a complete tail, or a long neck, or a good part of the trunk. Skulls are rare, unfortunately. Professor Osborn suggests that we have here the site of an old river-bar

[1] *The Century Magazine*, September, 1904. *The Fossil Wonders of the West*, by Professor H. F. Osborn.

HIND-LEG BONES OF DIPLODOCUS, *in situ*, BONE-CABIN QUARRY, MEDICINE BOW, WYOMING.

Reproduced by permission of Prof. Henry Fairfield Osborn and the American Museum of Natural History.

PLATE XX.

which arrested the floating carcases (more or less decomposed) which had slowly drifted down the stream towards it.

Diplodocus is almost the largest fossil reptile that has yet been discovered. From the illustration showing one of the hind limbs the reader probably surmises that Brontosaurus was a near relation, which is true. Certain bones, and a skull of the former, have been known for the last twenty years, but no *one* complete skeleton has ever been found. However, through the labours of Marsh, Hatcher, and Dr. W. J. Holland, enough material has been collected to build up a complete model or reconstruction

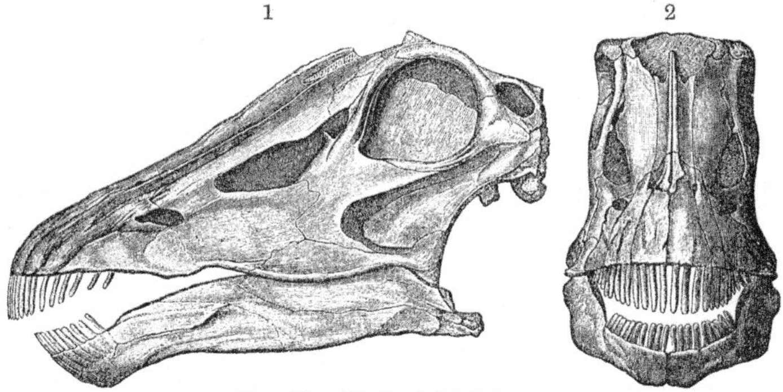

FIG. 52.—Skull of *Diplodocus*.
1. Side view. 2. Front view.

of the skeleton of this truly wonderful giant reptile, as shown in Plate XXI. It is a cast in plaster-of-Paris—the gift of Mr. Andrew Carnegie, who, by his generous gifts, is doing much to further scientific researches in various directions. This model when first set up in the gallery devoted (chiefly) to modern reptiles attracted a great deal of attention. Not unnaturally, for it is so stupendous. A reptile 84½ feet long was something quite new to the public. This model is an exact copy of the original composite skeleton, to be seen in the Carnegie Museum, Pittsburgh. Each bone was cast in a separate mould made from the original bone, so that for all practical purposes this model is as good as the original. It seems a pity that there is no room for it in the fossil reptile gallery, along with Cetiosaurus and other allies. Both geologists and naturalists are hoping that some day our Government will give a grant for the extension of the British Museum (Natural History) in Cromwell Road, and then we might have a spacious gallery entirely devoted to fossil reptiles. Failing that, there is still a hope that some rich man may come forward to supply the funds for this great improvement. The material for this reconstruction, made under the direction of Dr. W. J. Holland, director of the Carnegie Museum, Pittsburgh, comes from four incomplete skeletons: One was found by Dr. J. L. Wortman, in an expedition made in the year 1899, another by Mr. O. A. Peterson, in another expedition the following year. Both specimens came from the same quarry, in Sheep Creek, in Albany, County Wyoming, which had previously yielded so much material to Dr. Wortman. Later on another two, or more, turned up, each incomplete. No complete skull has yet been found: Marsh had two incomplete skulls, and the Carnegie Museum has another. The American Museum of Natural History has two others (see Fig. 52) also incomplete. In these circumstances a restoration had to be made from the material at hand, and missing parts supplied by a study of the skull of Brontosaurus, which is a near relation. This seems justifiable; and we may add that the tail, consisting of 70 vertebræ, has been found complete: a large part of the trunk also. So, with the neck, Diplodocus had a height of 14 feet, if it stood erect as shown in Plate XXII. Professor Tornier, however, thinks the legs were more or less bent. This view is partly expressed in our restoration. The teeth are small and weak, probably serving to collect succulent vegetation. The nostrils open quite in the top

PLASTER-CAST OF SKELETON OF A DINOSAURIAN LAND-REPTILE (*Diplodocus carnegii*), OVER 80 FEET IN LENGTH, FROM UPPER JURASSIC, WYOMING, U.S.A.

PLATE XXI.

(Gallery of Reptiles, Natural History Museum, London.)

of the head, a position which would be convenient to an air-breathing animal that spent most of its time in the water.

With regard to what may be called the architecture of this huge skeleton, Professor Osborn says: "The backbone is indeed a marvel. The fitness of the construction consists, like that of the American truss-bridge, in attaining the maximum of strength with the minimum of weight."

Morosaurus is an allied form, also from the Atlantosaurus beds of Wyoming, with a long neck and small head. But no complete skeleton has yet been found.

Unfortunately, there are at present no complete skeletons known of English Dinosaurs related to the American forms above described. But, since the English fossils were first in evidence by many years, and Marsh's discoveries have confirmed in a remarkable way conclusions drawn by Owen, Huxley, Hulke, and Seeley, and others from materials that were rather fragmentary, it may be worth while to give some account of these remains and the interpretations they have received.

Dr. Buckland, in his *Bridgewater Treatise*, 1836, referred to a limb-bone in the Oxford Museum, from the great Oölite formation near Woodstock, which was examined by Cuvier, and pronounced to have once belonged to a whale; also a very large rib, which seemed whale-like. In 1838 Professor Owen, when collecting materials for his famous *Report on the Fossil Reptiles of Great Britain*, inspected this remarkable limb-bone, and could not match it with any bones known among the whale tribe; and yet its structure, where exposed, was like that of the long bone (humerus) of the paddle of a whale. Later on, he abandoned the idea that it once belonged to a whale, and it was thought that the extinct animal in question might have been a reptile of the crocodilian order. In time, a fine series of limb-bones and vertebræ was added to the Oxford Museum by Professor Phillips (Dr. Buckland's successor at Oxford), who pronounced them to be Dinosaurian. The name "Cetiosaurus" (or Whale-lizard), originally given by Owen, was unfortunate, because there is really nothing whale-like about it, except a certain coarse texture of some of the bones.

In 1848 Dr. Buckland announced the discovery of another limb-bone (a femur), which Owen referred to Cetiosaurus; it was four feet three inches in length. Between 1868 and 1870, however, a considerable portion of a skeleton was discovered in the same formation at Kirtlington Station, near Oxford. These remains were the subject of careful examination by Professors Owen and Phillips. The femur this time was five feet four inches long. Their studies threw much light on the nature and habits of Cetiosaurus.

Of late years, Mr. Alfred N. Leeds, who has so zealously collected fossil bones from the Oxford clay near Eyebury (especially those of Plesiosauri, see p. 79) has brought together a large number of bones of the Cetiosaurus. These have been acquired by the British Museum (Natural History Department), and may be seen at South Kensington, where a large part of a skeleton may be seen mounted. It consists of most of the tail, the left hind limb, and the right fore limb. The height at the hip is ten feet six inches, and the total length probably was sixty feet. The skull is not known. Here again it is possible that the legs were more or less bent outwards as in lizards and crocodiles.

It is evident that Cetiosaurus was closely allied to the American Brontosaurus (p. 150); and so these earlier English discoveries have gained much in interest from the light thrown upon them by Professor Marsh's huge Saurian.

Another English Saurian of this group was the Ornithopsis,

THE GREATEST KNOWN DINOSAUR (*Diplodocus carnegii*).

PLATE XXII. Length 84½ feet.

from Wealden strata in the Isle of Wight, which has been the subject of careful study by Mr. Hulke and Professor Seeley. Their conclusions, based on the examination of separate portions of the skeleton (such as vertebræ), have been singularly confirmed by the discovery of Brontosaurus.

In Ornithopsis the vertebræ of the neck and back, though of great size, were remarkably light, and yet of great strength. One of the vertebræ of the back had a body, or centrum, ten inches long. Hoplosaurus and Pelorosaurus were evidently reptiles closely allied to the above types; but at present are so imperfectly known that we need not consider them here. On a recent visit to Berlin the writer was informed by Professor Brauer of the discovery in German East Africa of immense quantities of Dinosaurian bones. Among these one femur exceeds in length that of Diplodocus by several inches.

CHAPTER IX

DINOSAURS *(continued)*

"Everything in Nature is engaged in writing its own history: the planet and the pebble are attended by their shadows, the rolling rock leaves its furrows on the mountain side, the river its channel in the soil, the animal its bones in the stratum, the fern and the leaf inscribe their modest epitaphs on the coal, the falling drop sculptures its story on the sand and on the stone,—not a footstep on the snow or on the ground, but traces in characters more or less enduring the record of its progress."—EMERSON.

WE propose now to give a brief account of a wonderful group of Dinosaurs, first introduced to the scientific world through Dr. Mantell's labours. The first of these monsters is the Iguanodon, the earliest known individual of the "bird-footed" division (Ornithopoda). The history of the gradual reconstruction of its skeleton is an instructive instance of the results that may be obtained by a careful and patient study of fragmentary remains. Through the labours of Dr. Mantell, in the first half of the last century, a considerable knowledge was acquired of the greater part of the skeleton, but certain portions remained a puzzle; these, however, were eventually explained by Professor Huxley and Mr. Hulke, and a few years ago a series of complete skeletons was most fortunately obtained in Belgium, so that now every part of the huge framework of this monster is known to the palæontologist. Its history, as a fossil, is most interesting, and furnishes one more example of the marvellous insight into the nature of extinct animals displayed by the illustrious Baron Cuvier. Let us begin with the teeth, since they were the first part of the monster brought to light.

The first specimens of the teeth of Iguanodon were found by Mrs. Mantell, in 1822 (see Fig. 53), in the coarse conglomerate of certain strata in Tilgate Forest, belonging to the Cretaceous period (see Table of Strata, Appendix I.). Dr. and Mrs. Mantell subsequently collected a most interesting series of these remarkable teeth (which, for a time, puzzled the most learned men of

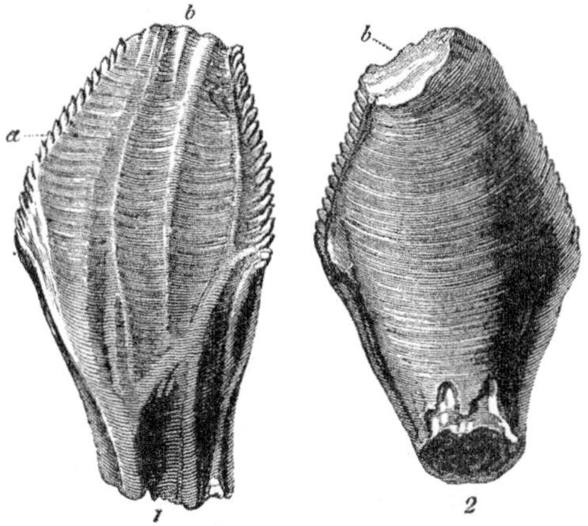

FIG. 53.—Tooth of *Iguanodon*, with the apex slightly worn. (From the Wealden Beds of Tilgate Forest. Natural size.) 1. Front aspect, showing the longitudinal ridges and serrated margins of the crown. 2. View of the back, or inner surface of the tooth. *a*. Serrated margins. *b*. Apex of the crown worn by use.

the day), from the perfect tooth of a young animal, to the last stage, that of a mere long stump worn away by mastication. In external form they bore a striking resemblance to the grinders of herbivorous mammals, and were wholly unlike any that had previously been known. The first specimen that arrested his attention was a large tooth, which, from the worn surface of its crown, had evidently once belonged to some herbivorous animal. In form it so entirely resembled the corresponding part of an incisor tooth of a large quadruped ground down by use, that Dr. Mantell was much embarrassed to account for its presence in the ancient Wealden strata, in which, according to all previous experience, no fossil remains of mammals would be likely to occur. No reptiles of the present day are capable of masticating their food; how, then, could he venture to assign it to a reptile? Here was a puzzle to be solved, and in his perplexity he determined to try whether the great naturalist at Paris would be able to throw any light on the question. Through Sir Charles (then Mr.) Lyell, this perplexing tooth was submitted to Baron Cuvier; and great was the doctor's astonishment on hearing that it had been without hesitation pronounced to be the upper incisor of a rhinoceros! The same tooth, with some other specimens, had already been exhibited at a meeting of the Geological Society, and shown to Dr. Buckland, Mr. Conybeare, and others, but with no more satisfactory result. Worse than that: Dr. Mantell was told that the teeth were of no particular interest, and that, without doubt, they either belonged to some large fish, or were the teeth of a mammal, and derived from some superficial deposit of the "glacial drift," then called Diluvium!

There was one man, however, who foresaw the importance of Mantell's discovery, and that was Dr. Wollaston. This distinguished philosopher, though not a naturalist, supported the doctor's idea that the teeth belonged to an unknown herbivorous reptile, and encouraged him to continue his researches.

The workmen in the quarry were stimulated by suitable rewards, and at length the doctor's efforts resulted in the discovery of teeth which displayed the curious serrated edges, and the entire form of the unused crown. Having forwarded specimens and drawings of these to Paris, Dr. Mantell went to

London, and ransacked all the drawers in the Hunterian Museum that contained jaws and teeth of reptiles, but without finding any that threw light on this subject. Fortunately, Mr. Samuel Stutchbury, then a young man, was present, and proposed to show him the skeleton of an Iguana, which he had himself prepared from a specimen that had long been immersed in spirits. And now the puzzle was in a fair way to being solved; for, to his great delight, the doctor found that the minute teeth of that reptile bore a closer resemblance in their general form to those from Tilgate Forest than any others he had ever seen.

In spite of this fortunate discovery, however, others remained obstinate and unconvinced; and it was not until he had collected a series of specimens, exhibiting various stages of the teeth, that the correctness of his opinion was admitted, either as to their true interpretation, or the age of the strata in which they were imbedded. And now there came good news from Paris. Cuvier, with the fresh material submitted to him, had boldly renounced his previous opinion, and gave the weight of his great authority to the view maintained by the discoverer of these teeth. In a letter to the doctor he said that such teeth were quite unknown to him, and that they belonged to some reptile. He suggested that they implied the existence of a *new animal*, a *herbivorous reptile.* Time would either confirm or disprove the idea, and in the mean time he advised Dr. Mantell to seek diligently for further evidence, and, if part of a jaw could be found with teeth adhering, he believed he could solve the problem. In his immortal work, *Ossemens Fossiles,* Cuvier generously admits his former mistake, and said he was entirely convinced of his error.

Baron Cuvier alone amongst the doctor's friends or correspondents was able to give any hint as to the character and probable relations of the animal to which the recently discovered teeth belonged. Being hampered by arduous professional duties in a provincial town, remote from museums and libraries, Dr. Mantell transmitted to the Royal Society figures and drawings of the specimens, and, at the suggestion of the Rev. W. D. Conybeare, adopted the name Iguanodon (Iguana-tooth) for the extinct reptile, a name which pointed to the resemblance of its teeth to those of the modern iguana, a land-lizard some three to five feet long, inhabiting many parts of America and the West Indies, and rarely met with north or south of the tropics.

In all living reptiles the insects or vegetables on which they feed are seized by the tongue or teeth, and swallowed whole, so that a movable covering to the jaws, similar to the lips and cheeks of the mammalia, is not necessary, either for seizing and retaining food or for subjecting it by muscular movements to the action of the teeth. It is the power of perfect mastication possessed by the Iguanodon that is so strange, for it implies a most remarkable approach in extinct reptiles to characters possessed now only by herbivorous mammalia, such as horses, cows, deer, etc. From this and other strange characters seen in the Dinosaurs, we learn that they in their day played the part of our modern mammals, whether carnivorous or herbivorous, and showed remarkable approach to the mammalian type, which of course is a much higher one. It is, therefore, not to be wondered at that Dr. Mantell's contemporaries, with the exception of Cuvier, found in these teeth an awkward puzzle and refused to believe that they belonged to a reptile. Such a notion was at variance with all previous experience; and we naturally form our conclusions largely by experience. The importance of discovering, if possible, a portion of the jaw of an Iguanodon was fully recognised by Dr. Mantell, and, urged by the encouragement he had received from the illustrious Cuvier, he eagerly sought for the required evidence. But nearly a quarter of a century elapsed before it was forthcoming. In the years 1841 and 1848 portions of the lower jaw with some attached teeth were found. It is very interesting in the light of recent discoveries to read the conclusions arrived at by Mantell and Owen, with regard to the organisation of this great Wealden reptile, and to see how, with the exception of certain details, they have been confirmed. Considering the imperfect nature of the materials at their command, it is wonderful that their forecasts should have turned out so successful. Thus Professor Owen predicted for the Iguanodon a total length of twenty-eight feet, and specimens discovered of late years show a length of twenty-four feet.

Dr. Mantell also was the first to prove, from the nature of the Wealden strata, that they were deposited in or near the estuary of a mighty river. With regard to the aspect of the country in which the Iguanodon flourished, he showed that coniferous trees probably clothed its Alpine regions; palms and arborescent ferns, and cycadaceous plants (*i.e.* plants resembling the modern zamia or "false palm") constituted the groves and forests of its plains and valleys; and in its fens and marshes the equisetacea (mare's tails) and plants of a like nature prevailed. Since the days of Dr. Mantell, the remains of Iguanodon or other closely allied genera have been found on the Continent, in England, and North America, in strata of various ages, from the Trias or New Red Sandstone to the Chalk (see Table of Strata, Appendix I.). The American Hadrosaurus must have been an allied form.

In the year 1878 was announced one of the most fortunate discoveries known in the whole history of geological science—a discovery unique of its kind, and one which throws considerable light on the nature of the monster first discovered by Dr. Mantell.

In that year came the good news that no less than twenty-three Iguanodons had been found in the colliery of Bernissart, in Belgium, between Mons and Tournai, near the French frontier. The specimens so skilfully set up by M. de Pauw represent two distinct species. The larger one, Iguanodon Bernissartensis, cannot be less than fifteen feet high, and, measured from the tip of the snout to the end of the tail, is rather over thirty feet long,

FIG. 54.—Skeleton of *Iguanodon Bernissartensis.* (After Dollo.)

covering nearly twenty-four feet of ground in its erect position (see Fig. 54). Iguanodon Mantelli is smaller and more slender looking, with a height of over ten feet, and a length of about twenty feet. (See Fig. 55.)

The huge three-toed impressions found in Sussex (see p. 166) prove that the monster, although owning a body as large as that of an elephant, habitually walked on its hind legs ! Some of the thigh-bones found by Dr. Mantell measured between four and five

IGUANODON BERNISSARTENSIS.

Length about 30 feet.

PLATE XXIII. Illustration lent by H. R. Knipe (from *Nebula to Man*).

feet in length. It will be seen that the fore limbs are small in comparison to the hind limbs. A remarkable feature of the hand is the large pointed bone at the end of the thumb, forming a kind of spur. The conical shape of this bone found by Dr. Mantell, who had no clue to its place in the skeleton, led him to suppose that it was a horn answering to that of a rhinoceros—a conclusion which Professor Owen refused for various reasons to accept. The latter concluded that it belonged to the hand, and now we see

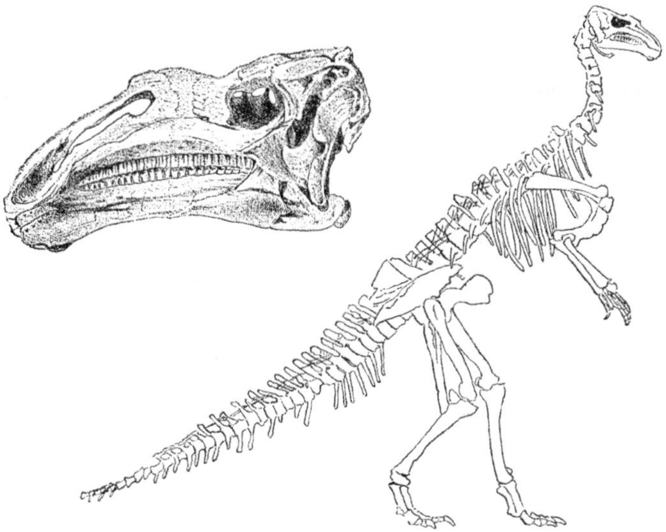

FIG. 55.—Skull and skeleton of *Iguanodon Mantelli*. (After Dollo.)

that he was right. It has been suggested that the spur was a weapon of offence, and that, when attacked, an Iguanodon may have seized its aggressor in its short arms, and made use of the spur as a dagger. But this is only conjecture, and perhaps the spur may have been useful in seizing and pulling down the foliage and branches of trees, or in grubbing them up by the roots. Detached specimens of this curious bone may be seen among the other remains of Iguanodon at South Kensington, and also some of the gigantic tracks already alluded to. A complete cast of a skeleton may be seen in the reptile gallery of the Natural History Museum.

The Bernissart specimens even afford some evidence as to the nature of the integument, or skin, and this supports the idea previously held that the creature possessed a smooth skin, or, at least, only slightly roughened. The muzzle was quite toothless, and perhaps may have been sheathed in horn, like the beak

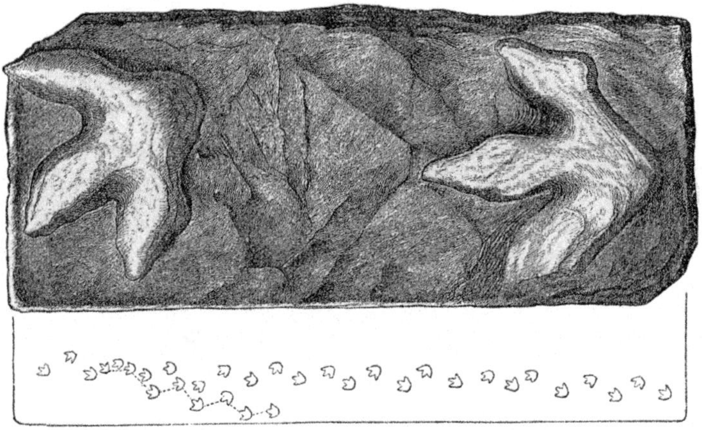

FIG. 56.—Tracks of *Iguanodon*, much reduced. From Wealden strata, Sussex.

of turtles—an arrangement highly useful for biting off the leaves of trees.

Probably it passed much of its time in the water, using its immense powerful tail as an organ of propulsion. When swimming slowly it may have used both sets of limbs, but when going fast it probably fixed its fore limbs closely beside its body, and drove itself through the water by means of the long hind limbs alone. M. Dollo, of Brussels, is preparing a final monograph on the Bernissart Iguanodons, a work to which palæontologists eagerly look forward. The Brussels Museum now contains ten mounted specimens, and twelve more lying in position on the rock. There cannot be much doubt that these unarmoured Dinosaurs were molested and preyed upon by their carnivorous contemporaries, such as the fierce Megalosaurus, previously described (p. 134). And with regard to this, M. Dollo makes the suggestion that, when on land, their great height and erect posture enabled them to descry such enemies a long way off. Their great height must also have stood them in good stead, by enabling them easily to reach the leaves of trees, tree-ferns, cycads, and other forms of vegetable life, which constituted their daily food. (See Plate XXIII.)

Should the reader visit the "geological island" in the grounds of the Crystal Palace, he will see that Mr. Waterhouse Hawkins's great model Iguanodon there set up is by no means in accordance with the description given above; but we must remember how imperfect was the material at his command. Mr. Karl Hagenbeck has recently added to his famous Zoological Gardens at Hamburg a number of life-size models of extinct animals, thus following in the footsteps of Mr. Waterhouse Hawkins, but with happier results. It is a step in the right direction, for the Science of Palæontology has made such great strides in the last forty or fifty years that a great deal of light has been thrown on the ancestry of living animals, and thus the student of life as it is to-day finds himself led back into past ages in order to get some glimpses into the evolution of the various orders and families now inhabiting the world.

Just as the Ceratosaurus of the New World is represented by, or rather corresponds to, the Megalosaurus of the Old World, so the other newly discovered Dinosaur, which we are about to describe, shows a good deal of correspondence with the now well-known Iguanodon. It is called by Professor Marsh, Claosaurus; and it represents a whole family of herbivorous Dinosaurs. In length it was nearly thirty feet, and must have stood nearly fifteen feet high. The remains of this remarkable creature were discovered by two of Professor Marsh's assistants in the Laramie

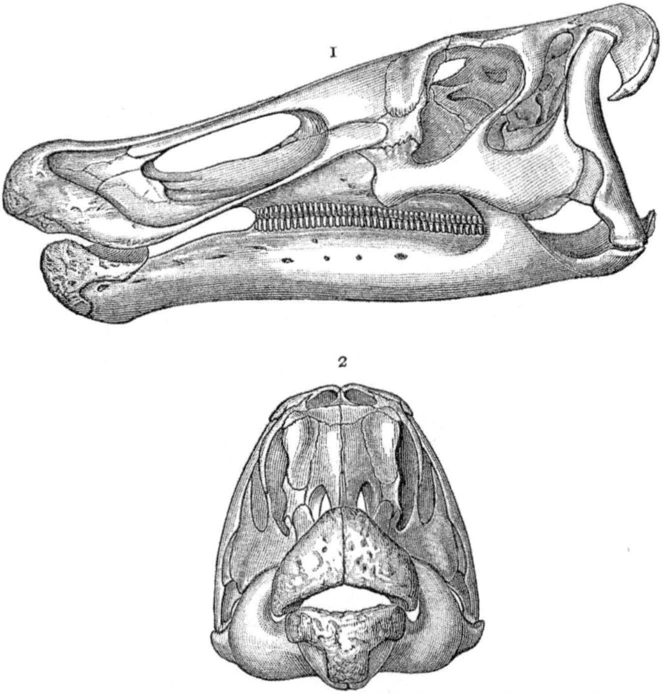

Fig. 57.—1. Skull of *Claosaurus annectens*, side view. 2. The same, front view. (After Marsh.)

beds of Wyoming, associated with those of the still more strange Triceratops, and also of some diminutive mammals lately discovered.

The most important feature in the skeleton of the Claosaurus is the skull. This is long and narrow, and, as seen in Fig. 57,

PLATE XXIV.

A LARGE HERBIVOROUS DINOSAUR, CLAOSAURUS, FROM NORTH AMERICA. CRETACEOUS PERIOD.

The bird represented on the right is a young *Hesperornis*.

has a large muzzle, the surface of which is considerably roughened, showing that it was covered, during life, with hard and thickened skin. The brain of the creature was very small in proportion to its skull. The orbits for the eyes are large, suggesting large eyes; and its sense of smell was probably pretty keen. The teeth resemble those of Hadrosaurus, an allied form. Succulent vegetation probably was the main diet of this animal. The whole backbone, or vertebral column, was found complete, with the

Fig. 58.—Skeleton of a large herbivorous Dinosaur, *Claosaurus annectens*, from Cretaceous strata, North America. (After Marsh.)

exception of a few little vertebræ from the end of the tail. Altogether there were about ninety vertebræ in the backbone of this powerful creature. The reader will perceive, from the drawing of the skeleton, in Fig. 58, that this Dinosaur must have possessed a long and powerful tail, such as would have constituted a most useful organ of propulsion through the water. As in its ally, Iguanodon, the vertebræ of the tail are compressed in a vertical direction. Plate XXIV., which is a restoration, should be compared with the drawing of the skeleton in Fig. 58. Another important feature of the skeleton is the remarkable ossified tendons, by means of which the neural spines of the vertebræ were held together, and so strengthened. They increase in number in the region of the thighs, and also in the first part of the tail.

The reader will have noticed the smallness of the fore limbs compared with the hind limbs. This is often the case with Dinosaurs. Only three fingers of the hand were used—the first was a rudimentary one; the second and third were of nearly equal length, the fourth shorter, and the fifth entirely wanting. These long fingers ended in hoofs, not claws; the fore limb as a whole was adapted for walking and support, rather than for purposes of prehension, although the latter might have been expected from its small size and position. The hind limb has three digits, all well developed and massive. These were covered with fairly broad hoofs, and it is clear that this Dinosaur was in the habit of walking on its hind legs. It is distinguished from Hadrosaurus, a near ally, by the fact that its limb-bones, instead of being hollow, are all solid. This piece of evidence tends to confirm the idea that Claosaurus was fond of the water; for it has been pointed out by Owen, Marsh, and other authorities, that land animals, having to support the weight of their bodies on land, have hollow bones; while those which live in the water, like whales, being buoyed up by the action of the water, and thus having so much weight taken off their hands and feet, have their limb-bones solid.

The late Professor von Zittel considered that possibly Claosaurus was not to be distinguished from Hadrosaurus.

There is an unusually fine example of a mounted skeleton in the United States National Museum (Washington), to which the name Trachodon has been given, but it is really the same genus. Professor F. A. Lucas says that in some cases the impression of the skin was preserved in the surrounding rock, and that it appears from these that the creature was covered with small horny plates. Another at Yale University is twenty-nine feet long. The Natural History Museum possesses a specimen showing the skin. Gallery IV., Wall-case 8.

The late Mr. J. W. Hulke, another English authority on Dinosaurs, described some twenty years ago a very remarkable and slender form of Dinosaur, on which we must now, in conclusion, say a few words. Professor Huxley has also helped to unravel the meaning of its structure. Hypsilophodon is the name this little creature has received, on account of the nature of its teeth. As far as we know it was the smallest of the Dinosaurs with the exception of the Compsognathus[1] and Hallopus, and Scleromochlus. It is certainly the least specialised—that is, the least highly organised. It lived during the Wealden period, or, in other words, during the closing scenes of the great Mesozoic era, and so was contemporary with others of a much more specialised character. As a rule the remains of animals found in our English Wealden and Cretaceous strata are in a fragmentary condition; but in this case we have, fortunately, an exception. Almost the whole of its skeleton is now known, thanks to the labours of Mr. Hulke and the Rev. William Fox. The remains occur in a bed which crops out a short distance west of Barns High Cliff, and passes under the shore a few yards west of Cowleaze Chine, on the south coast of the Isle of Wight. A restoration of its skeleton, according to Hulke, is seen in Fig. 59, and in Plate XXV. we have the creature restored to life. The skull shows a combination of characters belonging to a crocodile and a lizard, but is

[1] Greek—*kompsos*, elegant; *gnathos*, jaw.

on the whole more lizard-like. The creature was about four feet long.

The researches of Baron Nopcsa show that this Dinosaur was provided with a thin but well-developed armour, consisting of large, thin, and flat bony plates, feebly punctured. The base of the skull is very bird-like, as in some of the group Theropoda. The eyes had no sclerotic plates, as Mr. Hulke thought.

But there were in existence during the long Jurassic period other and even stranger forms of Dinosaurs with armour. One of these, only imperfectly known at present, was the many-spined Polacanthus.[1] This remarkable monster had the whole

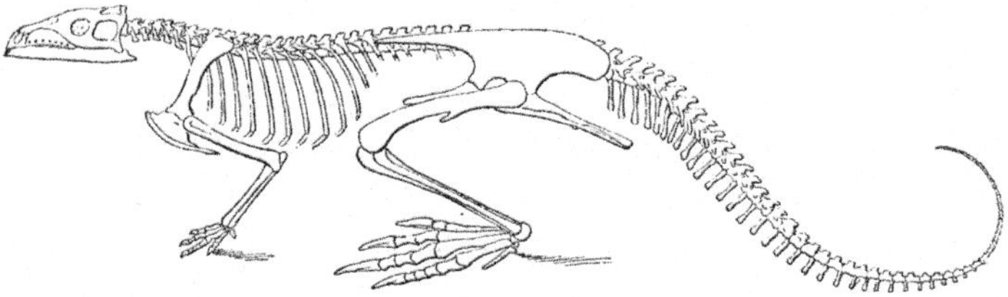

FIG. 59.—Skeleton of a small Dinosaur, *Hypsilophodon Foxi*, from Wealden strata, Isle of Wight. Restored by Professor J. W. Hulke, F.R.S. From the *Philosophical Transactions of the Royal Society*, vol. 173.

region of the loins and haunches protected by a continuous sheet of bony plate armour, rising into knobs and spines, after the fashion of the shield or carapace of certain extinct armadillos known as Glyptodonts (see Chap. XV.). Fig. 60 shows a partial reconstruction of the skeleton. The shaded parts of the drawing indicate the parts actually preserved, the rest is reconstructed with the help of the bones of an allied Dinosaur: *e.g.* Stegosaurus. From this we see that Polacanthus was a reptile of low stature. It was a slow-moving vegetable feeder, according to Baron Nopcsa,

[1] From Greek—*polus*, many, and *acantha*, spine.

A SMALL DINOSAUR, HYPSILOPHODON. PTERODACTYLS IN THE AIR. CRETACEOUS PERIOD.

PLATE XXV.

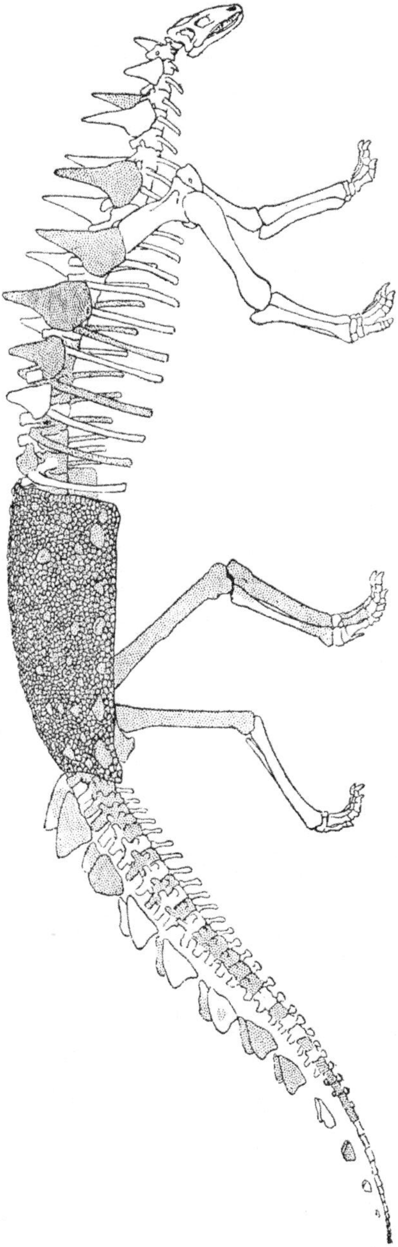

FIG. 60.—Reconstruction of skeleton of an armoured Dinosaur, *Polacanthus foxi*. (By Baron Nopcsa.) Wealden strata, Isle of Wight. Height about 3 feet. By permission of Baron Nopcsa and the *Geological Magazine*.

who, with the assistance of Dr. Smith Woodward and the late Mr. Barlow, made the above restoration which may be seen at South Kensington. Baron Nopcsa has made a special study of Dinosaurs in Hungary. Dr. Mantell had already pointed out certain analogies between Iguanodon and the huge extinct sloths of the South American continent, that flourished in the much more recent Pleistocene period; and this idea is now considerably strengthened by the later discoveries of armoured Dinosaurs. These are his words: "In fine, we have in the Iguanodon the type of the terrestrial herbivora which, in the remote epoch of the earth's physical history termed by geologists *the age of Reptiles*, occupied the same relative position in the scale of being, and fulfilled the same general purposes in the economy of nature, as the Mastodons, Mammoths, and the Mylodons (extinct sloths) of the Tertiary period, and the existing pachyderms."

It is, perhaps, one of the most interesting discoveries of modern geology, that certain races of animals now extinct have in various ways assumed some of the characteristics presented by animals much higher in the scale of being, that flourish in the present day. It seems as if there had been some strange law of anticipation at work, if we may venture so to formulate the idea. It has already been shown how the great saurians Ichthyosaurus and Plesiosaurus presumed to put on some of the characters of whales, and to play their *rôle* in nature, though they were only reptiles; how the carnivorous Dinosaurs acquired teeth like those now possessed by lions and tigers, which also are mammals; and now we find herbivorous Dinosaurs imitating the Glyptodon, an armadillo that lived in South America almost down to the human period. We shall not lose sight of this very interesting and curious discovery, for other cases will present themselves to our view in future chapters. The reader might ask, "If reptiles were able in these and other ways to imitate the mammals of to-day, or of yesterday, why should they not have been able to go a few steps further, and actually *become* mammals?" The Evolutionist, if confronted with such a question, would say, that there is no evidence of Dinosaurs turning into mammals, but that both may have branched off at an early geological period (say the Permian) from a primitive group of reptiles, or even of amphibians.

It must be borne in mind that, during the "age of reptiles" (Mesozoic period), the mammalian type was but feebly represented by certain small and humble forms, probably marsupials. As far as we know, there were no big quadrupeds such as flourish to-day; therefore reptiles played their part, and in so doing acquired some of their habits and structural peculiarities. It is difficult for us, living in an age of quadrupeds, to realise this, and to picture to ourselves reptilian types posing as "lords of creation," or, to use a homely phrase, "strutting in peacock's feathers."

We now pass on to describe two of the strangest and most wonderful of all the Dinosaurs, recently discovered in the far West. The first of these is the Stegosaurus,[1] or plated lizard, not wholly unknown before, because part of its skeleton was found some years ago in a brickfield in the Kimmeridge Clay at Swindon. It has been proved that some of the bones to which the name Omosaurus[2] has been applied really belonged to the former genus.

With such complete specimens now known by Professor Marsh's descriptions, it will not be necessary to mention the meagre remains discovered in this country, or the conclusions arrived at by Owen and Seeley, interesting as they are.

[1] Greek—*stegos*, roof of covering; *sauros*, lizard.
[2] Greek—*omos*, humerus, and *sauros*, lizard.

In the year 1877 Professor Marsh described, in the *American Journal of Science*, a considerable portion of a skeleton of a Stegosaur, remarking that this genus proved to be one of the most remarkable animals yet discovered. It was found on the eastern flank of the Rocky Mountains, in strata of Jurassic age; the bones indicated an animal about twenty-five feet long, and for this discovery Science is indebted to Professor A. Lakes and Engineer H. C. Beckwith of the United States Navy, who found the remains in Colorado, near the locality of the gigantic Atlantosaurus. The solid limb-bones seem to point to an aquatic life, but there can be little doubt that the monster did not pass all its time in the water. (Fig. 61 shows the skeleton.)

In 1879 Professor Marsh announced the discovery of additional remains from several localities. The most striking feature—from which the Stegosaur takes its name—was the presence of huge bony plates belonging to its skin, as well as large and small spines. Some of the plates were from two to three feet in diameter, and they were of various shapes. Of the spines, some were of great size and power, one pair being each over two feet long! The skull was remarkably small, and more like that of a lizard than we find in most Dinosaurs; the jaws were short and massive. Little was known at first of the brain, but fortunately a later discovery showed the brain-case well preserved. Later still, more than twenty other specimens of the Dinosaur were obtained, so that nearly every portion of the skeleton is now known. The skulls indicate that the creature possessed large eyes and a considerable power of smell. The jaws contain but a single row of teeth in actual use; but as these wore out, they were replaced by others lodged in a cavity below. Teeth, however, were not its strong point; they indicate a diet of soft succulent vegetation. The vertebræ have the ends of their centra more or less bi-concave. Many curious features in the skeleton can only be explained with reference to the heavy

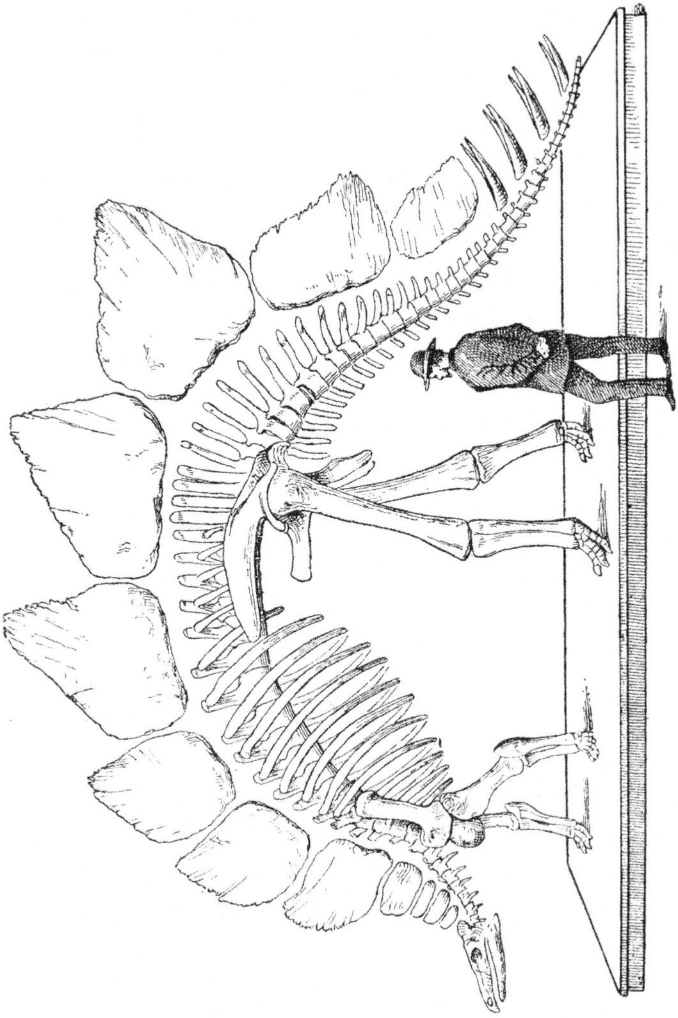

FIG. 61.—Skeleton of *Stegosaurus ungulatus*; length about 25 feet. (After Marsh.)

armour of plates and spines with which the Stegosaur was

provided. Thus the vertebræ have their "neural spines" expanded at the summit to aid in supporting part of the armour. (See Fig. 62.) The fore limbs were short and massive, but provided with five fingers; the hind limbs were very much larger and more powerful. These and the powerful tail show that the monster could support itself on them as on a tripod, in an upright position, and this position must have been easily assumed in consequence of the massive hind quarters. As in Iguanodon, there were three toes to the hind feet, and these were probably

1 2

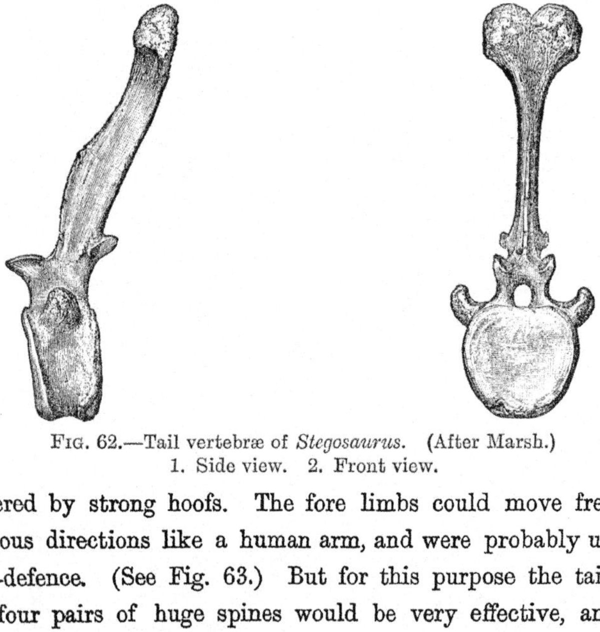

FIG. 62.—Tail vertebræ of *Stegosaurus*. (After Marsh.)
1. Side view. 2. Front view.

covered by strong hoofs. The fore limbs could move freely in various directions like a human arm, and were probably used in self-defence. (See Fig. 63.) But for this purpose the tail with its four pairs of huge spines would be very effective, and one could easily imagine that a single deadly blow from such a tail would be sufficient to drive away, if not kill, one of the carnivorous enemies of the species. All the plates and spines were, during life, protected by a thick horny covering, which

Plate XXVI.

STEGOSAURUS UNGULATUS. FROM STRATA OF JURASSIC AGE. NORTH AMERICA.
Length about 25 feet.
Illustration lent by H. R. Knipe (from *Nebula to Man*).

must have increased their size and weight. Such a covering seems to be clearly indicated by certain grooves and impressions that mark their surfaces. (See Fig. 64.) The largest plates are

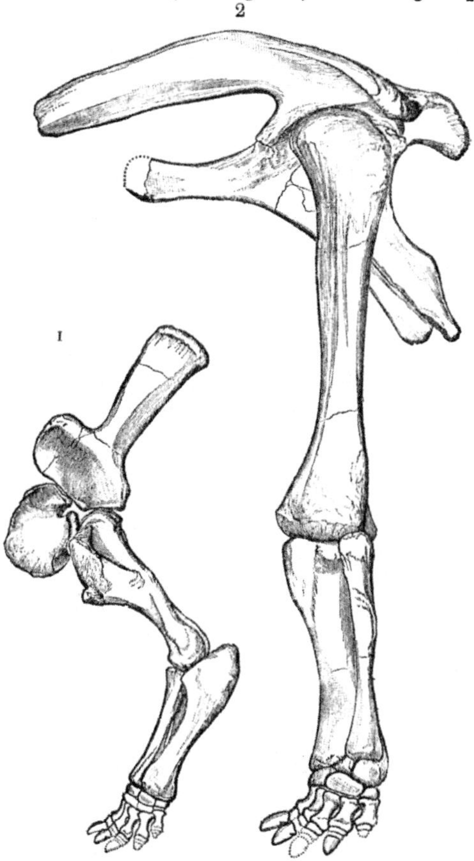

FIG. 63.—Limb-bones of *Stegosaurus*. (After Marsh.)
1. Fore leg. 2. Hind leg.

unsymmetrical, and were probably arranged along the back, in rows, as in our restoration, Plate XXVI. It will be noticed, by those who are familiar with our first edition, that this Plate gives a somewhat different representation of the Stegosaur, in which the length of the hind limbs is more apparent, and also they are more free from the body.

Finally, the Stegosaur displays a rather remarkable feature; for a very large chamber was found in the sacrum [1] formed by

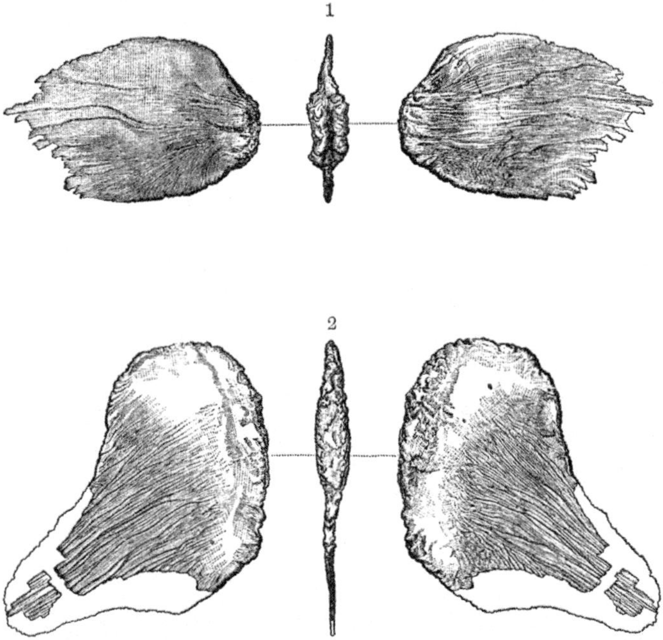

FIG. 64.—1, 2. Plates of Stegosaurus. The middle figures show their thickness. (After Marsh.)

an enlargement of the spinal cord.

The next monster to be described is one that has fortunately left to posterity a good record of itself, and probably was not very unlike the Hylæosaurus of Mantell. This is the

[1] The sacrum may be thus defined: the Vertebræ (usually fused together) which unite with the haunch-bones (*ilia*) to form the pelvis.

Scelidosaurus: so named by Professor Owen from the indications of greater power in the hind legs than in most saurians.[1] It is the only known example of an almost entire skeleton of an English Dinosaur, and the history of its discovery is rather curious. Some time previous to 1861, Mr. J. Harrison, of Charmouth, obtained from the Lower Lias of that neighbourhood portions of the hind limb of a Dinosaur, and, later on, a nearly complete skull. These specimens were described by Owen, and the genus was founded on them. Mr. Harrison, whose discovery aroused great interest, continued to search on the same spot, and was rewarded by finding all the rest of the skeleton, except most of the neck vertebræ. This was extracted in several blocks, and these, after careful "development" of the bones, were fitted together so as to exhibit the whole skeleton. This most valuable specimen can now be seen at South Kensington in a separate glass case, and is one of the treasures of the unrivalled gallery of fossil reptiles. (The case is placed so that both sides of the specimen can be seen.) Its length is about twelve feet; perhaps the individual it represents was not fully grown, but, on account of the absence of most of the neck vertebræ, it is impossible to give the exact length. Both hind limbs are entire and well seen, but of the fore limbs the hands are wanting. The former were provided with four "functional" toes—that is, toes that were used,—and one "rudimentary" or unused one. There were two big spines, one placed on each shoulder, and a series of long plates arranged in lines along the back and side.

Another Dinosaur, of considerable dimensions, which flourished during the Wealden period was the Hylæosaurus, discovered by Dr. Mantell, and so named by him because it came from the Weald.

[1] From Greek—*scelis*, limb, and *sauros*, lizard.

Between 1832 and 1837 he discovered three specimens showing ribs, limb-bones, vertebræ (of the tail), bony plates, and spines. Unfortunately the skull and teeth are unknown; but the tail appears to have been nearly six feet long. Mr. Waterhouse Hawkins's model in the grounds of the Crystal Palace, at Sydenham, near that of the Iguanodon, is rather a premature attempt at restoration.

The last, and in some ways the strangest of the Dinosaurs, was the Triceratops[1] that flourished in America at the end of the

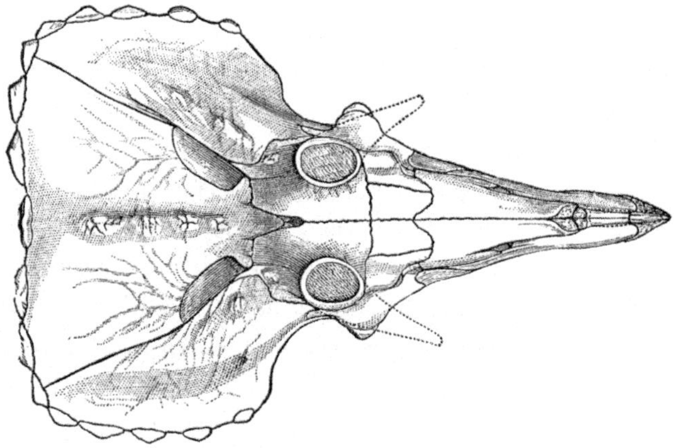

FIG. 65.—Skull of *Sterrolophus* allied to *Triceratops*, seen from above. (After Marsh.)

long Mesozoic era, during the Cretaceous period. The name refers to the three horn-cores found on the skull, which probably supported true horns like those of oxen. Whereas the Stegosaur was provided with quite a small skull, this monster had one of huge dimensions and remarkable shape (see Fig. 65 and Plates XXVII. and XXVIII.). In the younger ones it was about six feet long, but in an old individual must have reached a length of seven or eight feet. Such a skull is only surpassed by some whales of

[1] Greek—*treis*, three; *ceras*, horn; *ops*, face.

1. THE LATE PROF. O. C. MARSH.

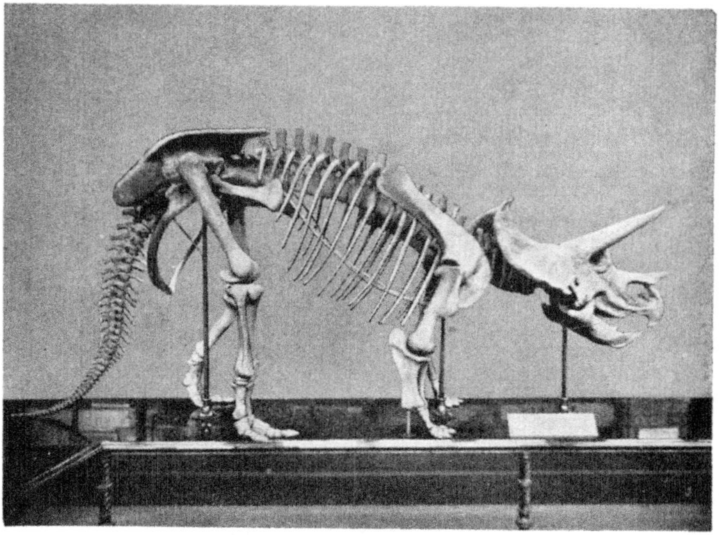

2. RESTORED SKELETON OF TRICERATOPS.

PLATE XXVII.

the present day. Twenty different skulls of this kind have been found, and Professor Marsh places the horned Dinosaurs in a separate family, to which he has given the name Ceratopsidæ, or horn-faced. Their remains come from the Laramie beds, believed to be of Cretaceous age, but representing a remarkably mixed fauna and flora, so that some have considered them to be Tertiary. The strata containing these fossils are very rich in organic remains, and have yielded not only other Dinosaurs, but Plesiosaurs, crocodiles, turtles, many small reptiles, a few birds, fishes, and small mammals. The Ceratops beds are of fresh-water or brackish origin, and can now be traced for nearly eight hundred miles along the east flank of the Rocky Mountains.

In this Dinosaur we find the fore feet larger than usual in proportion to the hind limbs, and there can be no doubt that it walked on all fours. Its length was about 19 feet 8 inches. The original estimate of Professor Marsh was about 25 feet, but Mr. Gilmore, who made the model shown in Plates XXVII. and XXVIII., has taken out six vertebræ so that the length is now reduced to 19 feet 8 inches. All the vertebræ and limb-bones are solid. The brain was smaller in proportion to the skull than in any known vertebrate.

The teeth are remarkable in having two distinct roots. The wedge-like form of the skull is also very peculiar. The two large horns come immediately over the eyes, and the small one above the nose; this Dinosaur was, therefore, well provided with weapons of offence, such as would be highly useful in driving away or wounding carnivorous enemies. The back part of the skull rises up into a kind of huge crest, and this during life was protected by a special fringe of bony plates. Such an arrangement doubtless formed an effective shield to ward off blows when one Triceratops was fighting another, as bulls or buffaloes of the present day fight with their horns. The mouths of these Dinosaurs formed a kind of beak, sheathed in horn.

The body as well as the skull was protected, but the nature and position of the defensive parts in different forms cannot yet be determined with certainty. Various spines, bones, and plates have been found that evidently were meant for the protection of the creature's body, and belonged to the skin. Probably some of these were placed on the back, behind the crest of the skull;

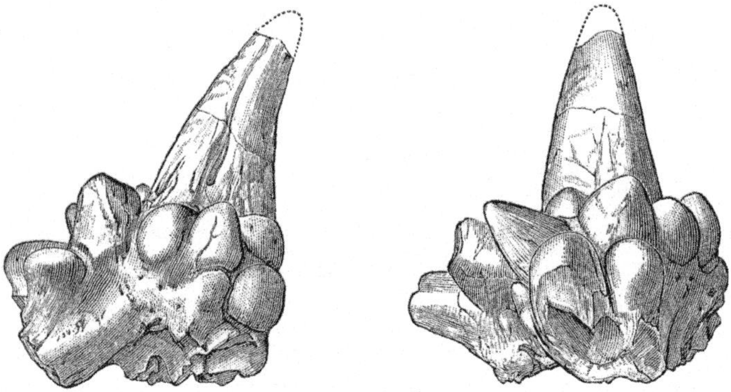

FIG. 66.—Bony spines possibly belonging to the skin of *Triceratops*. (After Marsh.)[1]

some may have defended the throat, as in Stegosaurus. Altogether, Triceratops is very different from any other Dinosaur. One cannot help picturing it rather as a fierce rhinoceros-like animal. In the restoration (Plate XXIX., Frontispiece) our artist has given it a thick skin, rather like that of the rhinoceros, only indicating small bony plates, etc., here and there.

Professor Marsh thinks that as the head increased in size to

[1] There is some doubt about these spines; for they might possibly belong to the carnivorous Dinosaur Tyrannosaurus found in the same place (see p. 139). Mr. Barnum Brown has recently described (1906) two sternal bones, previously unknown, but they are unlike those of Diplodocus and Brontosaurus.

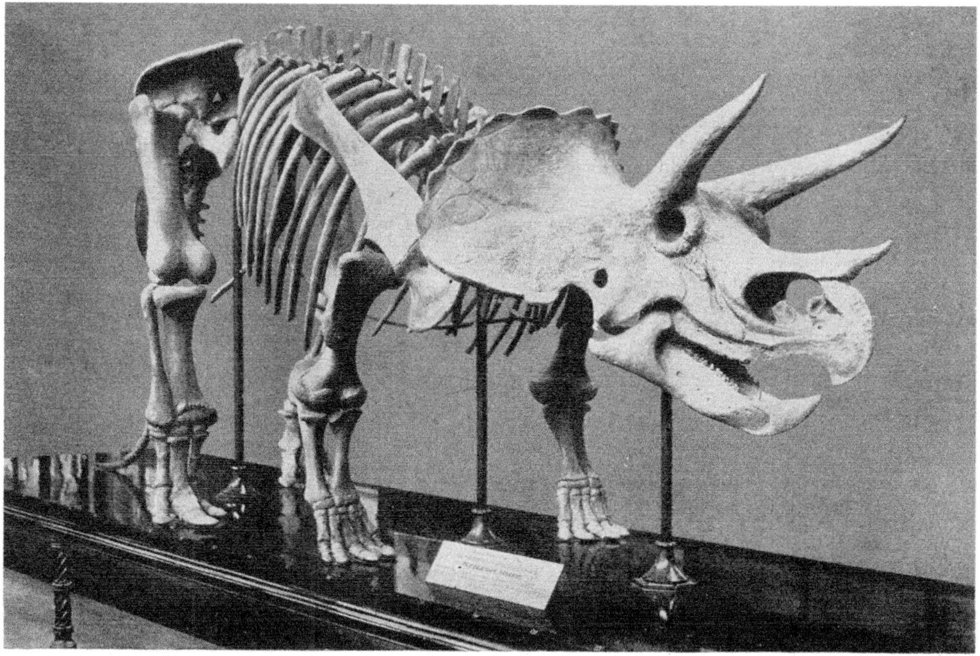

RECONSTRUCTED SKELETON OF A GREAT HERBIVOROUS HORNED DINOSAUR, TRICERATOPS PRORSUS, AS NOW MOUNTED IN THE NATURAL HISTORY MUSEUM, LONDON. CRETACEOUS PERIOD. NORTH AMERICA.

PLATE XXVIII.

bear its armour of bony plates, the neck first, then the fore feet, and then the whole skeleton was specially modified to support it; and he concludes that as these changes took place in the course of the evolution of this wonderful Dinosaur, the head at last became so large and heavy that it must have been too much for the body to bear, and so have led to its destruction! This conclusion, if sound, is a warning against carrying "specialisation" too far. If we wished to write an epitaph on the tomb of the monster, it ought (according to Professor Marsh) to be, "I and my race died of over-specialisation."

After all these various efforts to improve themselves and to perfect their organisation so as to bring it into harmony with their surroundings, or "environment," as the biologists say, it seems rather hard that the Dinosaurs should have been extinguished, and their place in Nature taken by a higher type; but all things have their day, even Dinosaurs.

With regard to the difficulties, hardships, and dangers attending the discovery and transport of the remains, Professor Marsh's concluding remarks may be quoted here, since they give us a glimpse into the nature of his explorations in the far West that have now become so famous. He says, "In conclusion, let me say a word as to how the discoveries here recorded have been accomplished. The main credit for the work justly belongs to my able assistant, Mr. J. B. Hatcher, who has done so much to bring to light the ancient life of the Rocky Mountain regions. I can only claim to have shared a few of the dangers and hardships with him, but without his skill little would have been accomplished. If you will bear in mind that two of the skulls weighed nearly two tons each, when partially freed from their matrix and ready for shipment, in a deep desert cañon, fifty miles from a railway, you will appreciate one of the mechanical difficulties overcome.

When I add that some of the most interesting discoveries were made in the hunting-grounds of the hostile Sioux Indians, who regard such explorations with superstitious dread, you will understand another phase of the problem. I might speak of even greater difficulties and dangers, but the results attained repay all past efforts, and I hope at no distant day to have something more of interest to lay before you." [1]

[1] *American Journal of Science*, vol. xli. p. 176.

CHAPTER X

SEA-SERPENTS

"Sand-strewn caverns, cool and deep,
Where the winds are all asleep;
Where the spent lights quiver and gleam;
Where the salt weed sways in the stream;
Where the sea-beasts, ranged all round,
Feed in the ooze of their pasture ground;
Where the sea-snakes coil and twine,
Dry their mail, and bask in the brine."

M. ARNOLD, *The Forsaken Merman.*

IT has been said that everything on earth has its double in the water. Are there not water-beetles, water-scorpions, water-rats, water-snakes, sea-lions, sea-horses, and a host of other living things, whether plants or animals, bearing some sort of resemblance to others that live on land? Then why not sea-serpents? The great controversy of the sea-serpent, that has so often been discussed in the newspapers, need not be considered here. We are dealing not with the present, but with the past; and whether or no the wonderful sailors' yarns of sea-serpents can be regarded as authentic, even in a single case, we can offer our readers infallible proof that during the so-called "Age of Reptiles," certain monstrous saurian animals flourished in considerable abundance, which, though not true serpents, nevertheless must have borne a striking resemblance to such, as they cleaved the waters of primæval seas. (See Appendix II.)

The modern evolutionist believes that snakes are descended

from lizards, possessing, as usual, four legs; that some primitive form of lizard with very small legs appeared on the scene, and found that it could better move along by wriggling its body and pushing with its ribs than by walking. So, in course of time, a race of lizards without legs arose; these, by Natural Selection, and perhaps other means, became more and more elongated, so that they could move faster than their ancestors, and glide out of harm's way more effectually. Thus was the snake evolved from a lizard.

Now, in the great geological museum of the stratified rocks, there have been discovered skeletons of marine reptiles, which propelled themselves chiefly by means of their tails and elongated bodies, rather than by their limbs. The limbs were not discarded entirely as in the case of the serpents, but were useful in their way as the fins of fishes are. Perhaps, therefore, we may be justified in calling these ancient monsters sea-serpents, in consideration of their long thin bodies; for they certainly would be called by that name if now living, and their skulls are very much like those of a modern lizard known as Varanus.

Strictly speaking, they were not serpents, but more or less like some of the extinct saurians described in Chap. V. The name, however, has been adopted by geologists, and is useful in so far as it serves to remind us of their very peculiar shape and structure. Remains of these strange creatures have been found both in Europe, North and South America, and New Zealand.

One of the earliest discoveries of remains of a fossil sea-serpent was made by M. Hoffmann, a Dutch military surgeon, in the year 1770. Maestricht, a city in the interior of the Netherlands, situated in the valley of the Meuse, stands on certain strata of limestone and sandstone, belonging to the Upper Chalk. Extensive quarries have, for many centuries, been worked in the sandstone, especially in the eminence called St. Peter's Mount, which is a cape or headland between the Meuse and the Jaar. This elevated plateau extends for some distance towards Liége, and presents an almost perpendicular cliff towards the Meuse. From the extensive works that have so long been carried on, immense quantities of stone have been removed, and the centre of the mountain is traversed by galleries, and hollowed by vast excavations. Innumerable fossils, such as marine shells, corals, crustaceans, bones and teeth of fishes, have been obtained from this rock, but St. Peter's Mount is now chiefly celebrated for the discovery of the bones and teeth of a huge saurian, to which Mr. Conybeare has given the name Mosasaurus, on account of its connection with the River Meuse. M. Hoffmann had long been an assiduous collector of fossils from this neighbourhood, and he had the good fortune to obtain the famous specimen on which this genus is founded.

It was at first considered by M. Faujas St. Fond to be a crocodile; but Cuvier and Camper formed a different and better conclusion. Perhaps no fossil ever had such a remarkable history as this one, as the following account, from M. Faujas St. Fond's work on the fossils of St. Peter's Mount,[1] will show.

"Some workmen, on blasting the rock in one of the caverns of the interior of the mountain, perceived, to their astonishment, the jaws of a large animal attached to the roof of the chasm. The discovery was immediately made known to M. Hoffmann, who repaired to the spot, and for weeks presided over the arduous task of separating the mass of stone containing these remains from the surrounding rock. His labours were rewarded by the successful extrication of the specimen, which he conveyed in

[1] *Histoire Naturelle de la Montagne de St. Pierre.* This account is given by Dr. Mantell, in his *Petrifactions and their Teaching*, 1851.

triumph to his house. This extraordinary discovery, however, soon became the subject of general conversation, and excited so much interest, that the canon of the cathedral which stands on the mountain resolved to claim the fossil, in right of being lord of the manor; and succeeded, after a long and harassing lawsuit, in obtaining this precious relic. It remained for years in his possession, and Hoffmann died without regaining his treasure, or receiving any compensation. At length the French Revolution broke out, and the armies of the Republic advanced to the gates of Maestricht. The town was bombarded; but, at the suggestion of the committee of savans who accompanied the French troops to select their share of the plunder, the artillery was not suffered to play on that part of the city in which the celebrated fossil was known to be preserved. In the mean time, the Canon of St. Peter's, shrewdly suspecting the reason why such peculiar favour was shown to his residence, removed the specimen, and concealed it in a vault; but when the city was taken, the French authorities compelled him to give up his ill-gotten prize, which was immediately transmitted to the Jardin des Plantes, at Paris, where it still forms one of the most striking objects in that magnificent collection."

Dr. Mantell quotes the Frenchman's remarks on this transaction: "*La Justice, quoique tardive, arrive enfin avec le temps:*" but adds, "The reader will probably think that, although the reverend canon was justly despoiled of his ill-gotten treasure, the French commissioners were but very equivocal representatives of *Justice!*"

The beautiful cast (Fig. 67) at South Kensington (Fossil Reptile Gallery III., Wall-case 1) was presented to Dr. Mantell by Baron Cuvier in 1825. It consists of both jaws, with numerous teeth, and some other parts (see Fig. 68). The length is about

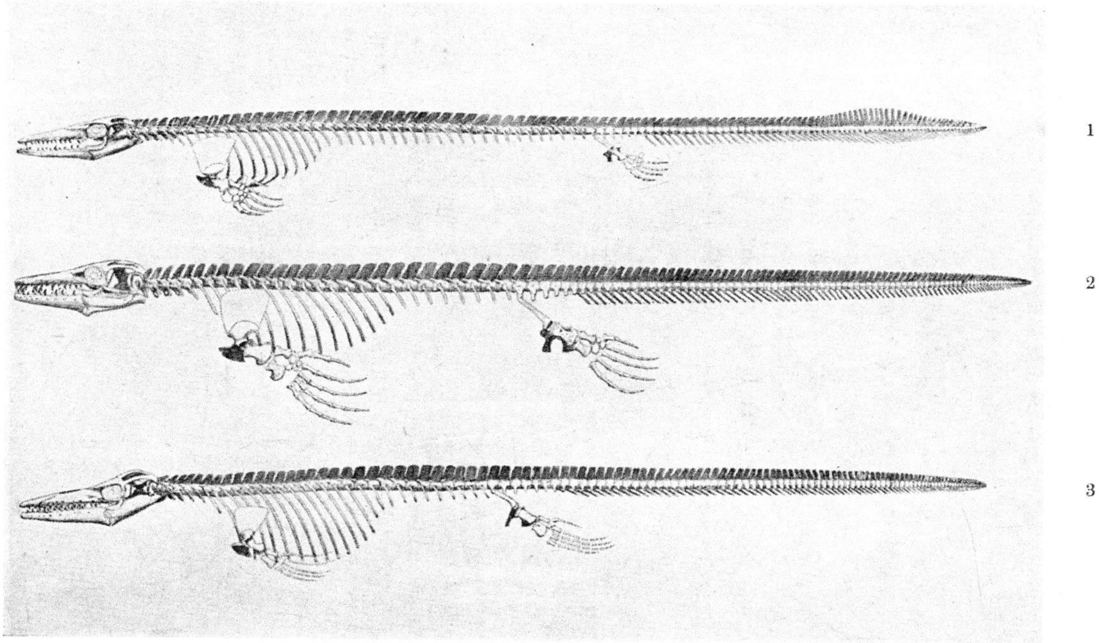

SKELETONS OF MOSASAUROID REPTILES FROM UPPER CRETACEOUS STRATA OF KANSAS, U.S.A., AFTER DR. S. W. WILLISTON.

PLATE XXX. 1. *Clidastes.* 2. *Platycarpus.* 3. *Tylosaurus.*

four and a half feet. This nearly perfect head was for a time a stumbling-block to many naturalists, some of whom were of

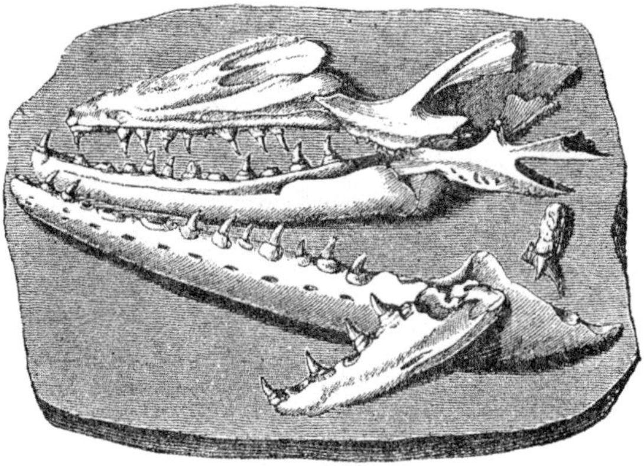

FIG. 67.—Skull of *Mosasaurus Hoffmanni*. The original is 4½ ft. by 2½ ft.

opinion that it belonged to a whale. Cuvier and others considered it to be a kind of link between the Iguanas and the Monitors.[1]

The entire backbone of the Maestricht animal appears to have consisted of one hundred and thirty-one vertebræ, of which ninety-seven belonged to the tail. The total length of the skeleton is estimated at twenty-four feet, and the head was about one-sixth of the total length. The tail is only ten feet long, whereas in a crocodile the tail exceeds the length of the body. Although in his day the limbs of the Mosasaurus were imperfectly known, Cuvier

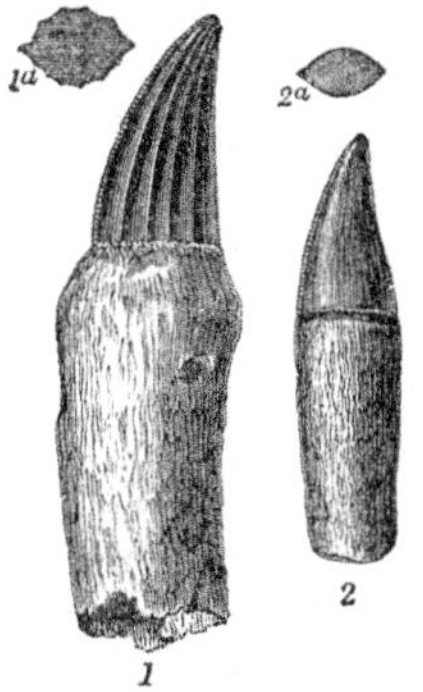

FIG. 68.—Teeth of *Mosasaurus* (½ natural size). 1a, 2a, transverse sections of the teeth.

[1] The Monitors are a family of large lizards inhabiting the warmer parts of Africa and Asia. They live near the banks of rivers, and some are altogether aquatic. They often devour the eggs of crocodiles and aquatic birds. The Nile Monitor, or Varanus, grows to a length of six feet.

rightly considered them to be adapted for swimming, and, with his usual foresight, concluded that this monster was a marine reptile of great strength and activity, having a large tail flattened vertically and capable of being moved from side to side with such force and rapidity as to be a powerful organ of propulsion, capable of stemming the most agitated waters. The large conical recurved teeth, the largest of which was nearly three inches long, are well seen in Figs. 67 and 68. Dr. Mantell was fortunate enough to find, in the year 1820, some vertebræ from the English Chalk, near Lewes, which were identified as belonging to a Mosasaurus. It is now known that this reptile attained the length of forty-five feet.

In 1831 a portion of a lower jaw with large conical teeth was discovered in the Chalk near Norwich. But these teeth were not quite similar to those of the Maestricht specimen, and Sir Richard Owen therefore founded upon them the new genus Leiodon.[1] But Leïodon must have been very similar to Mosasaurus.

Of late years many fine specimens have been discovered in North America, and the labours of Leidy, Marsh, and Cope have been of the greatest service in completing our knowledge of this strange group of saurians. In the American Cretaceous seas they ruled supreme, as their numbers, size, and carnivorous habits enabled them easily to vanquish all rivals. Probably some of them were seventy-five feet in length, the smallest being ten or twelve feet long. In the inland Cretaceous sea from which the Rocky Mountains were beginning to emerge, these ancient sea-serpents abounded; and many were entombed in its muddy deposits. On one occasion, as Professor Marsh rode through a valley washed out of this old ocean bed, he observed

[1] Greek—*leios*, smooth, and *odous*, tooth.

no less than seven different skeletons of these monsters in sight at once! The same authority mentions that the Museum of Yale College contains remains of not less than 1400 distinct individuals. In some of these the skeleton is nearly if not quite complete; so that every part of its structure can be determined with almost absolute certainty.

According to the late Professor Cope of Pennsylvania University, who made a special study of this group of extinct saurians, fifty-

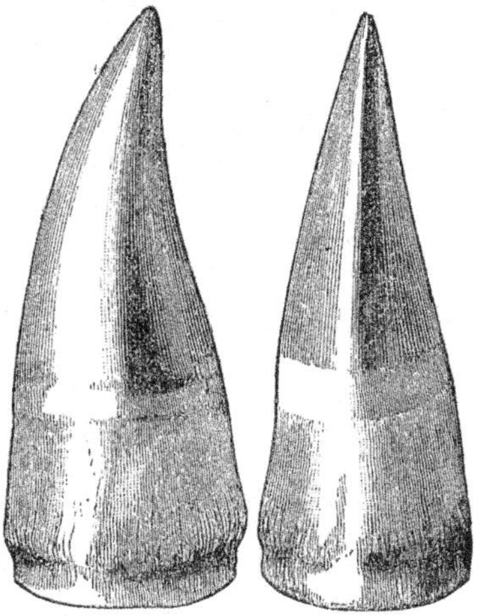

1 2

FIG. 69.—Lower tooth of *Leiodon*. 1. Side view. 2. Profile.

one species have been discovered in North America, in the States of New Jersey, Alabama, Kansas, North Carolina, Mississippi, and Nebraska. The same authority has shown that they were characterised by a wonderful elongation of form, especially of the tail; that their heads were large, flat, and conical in shape, with eyes directed partly upward; that they were furnished with two pairs of paddles like the flippers of a whale. With these flippers, and the eel-like strokes of their flattened tail, they swam with considerable speed. Like snakes, they were furnished with four rows of formidable teeth on the roof of the mouth, which served admirably for seizing their prey.

But the most remarkable feature in these creatures was the arrangement for permitting them to swallow their prey whole, in the manner of snakes. Thus each half of the lower jaw was articulated at a point nearly midway between the ear and the chin, so as to greatly widen the space between the jaws, and Professor Cope thinks that the throat must consequently have been loose and baggy.

Professor Cope, however, in giving the name Pythonomorpha to this ancient group, has pressed his views too far, and dwelt unduly on their supposed relationship with serpents. Other authorities regard them as essentially swimming lizards, with four well-developed paddles; and this is probably the right view to take of them.

"If, as appears certain," says the Professor, "the Mosasauroid, discovered by Webb, measures seventy-five feet in length, and the M. maximus, eighty, the Leiodon dyspelor must have been the longest reptile known, and approaches very nearly the extreme of the mammalian growth seen in the whales, though, of course, without their bulk. Such monsters may well excite our surprise, as well as our curiosity, in the inquiry as to their source of food-supply, and the character of those contemporary animals preserved on the same geological horizon."

According to Dr. S. W. Williston, who has, of late years, discovered and described most valuable specimens of Mosasauroid reptiles, Clidastes had a skull less flattened than that shown in some earlier restorations (*e.g.* in the previous edition of this book) and the digits were curved, as in Platycarpus. Of the former reptile, which was forty feet long, a single specimen was discovered by Dr. Williston, complete in nearly every detail. The restoration of the skeleton of Platycarpus, shown in Plate XXX., is based chiefly on one specimen, but certain parts are supplied conjecturally from Tylosaurus. The latter has been restored, as to its skeleton, from three specimens found by Dr. Williston, who was fortunate enough to find an impression of the skin, near one of the paddles, showing small scales, rather like those of a snake, but these were too small to be shown in our restoration (Plate XXXI.) of Tylosaurus. Platycarpus seems to be intermediate between Clidastes and Tylosaurus. In Clidastes we see a very long body, and a shorter tail, but this tail must have

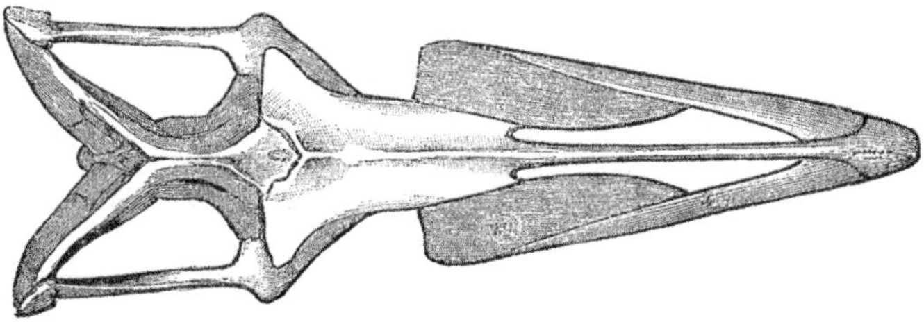

FIG. 70.—Superior aspect of the cranium of *Platycarpus curtirostris*, Cope; from the Upper Cretaceous of N. America (greatly reduced). (After Cope.)

been a powerful organ of propulsion. The limbs were strong and flexible, and the movements of this Mosasauroid reptile were more snake-like. The vertebræ in the tail had their neural arches lengthened, and probably the tail had a fin-like expansion to help in propulsion, as in the case of eels.

The accounts given by Professor Cope of his explorations and the difficulties encountered in procuring the valuable specimens on which his conclusions are based, are most interesting, and such as every fossil hunter will appreciate. We, in England, who visit clay pits, stone quarries, railway cuttings, etc., during a morning or afternoon walk, and return home at our leisure with a few small specimens in our pocket, or in a bag at our back, can hardly realise how arduous must be the work of finding, digging out, and transporting for such long distances, the remains of the monsters of Kansas, and other parts of North America.[1]

Leiodon proriger (Cope) was abundant in the old North American Cretaceous sea, and reached a length of seventy-five feet. It had a long projecting muzzle, somewhat like the snout of a sturgeon. Platycarpus and Tylosaurus had peculiarly sharp-pointed heads (see Figs. 69, 70, and 71).

A few words may be added here with regard to Professor Cope's important discovery of Leiodon—a genus already alluded

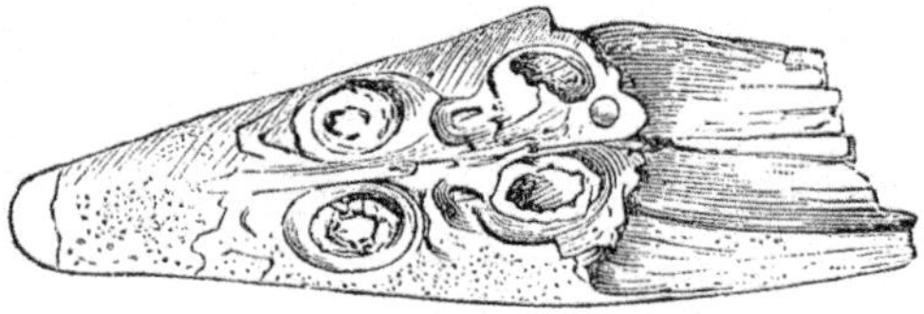

FIG. 71.—Snout of *Tylosaurus*, palatal view. (After Marsh.)

to as having been founded by Sir Richard Owen. The type specimen of Leiodon dyspelor,[2] which first indicated the characters of this wonderful species, was obtained from the yellow beds of the Niobrara epoch of the Jornada del Muerto, near Fort McRae, New Mexico. The greater part of the remains have been described by Professor Leidy. But a

[1] Mr. Charles H. Sternberg, who has spent his life collecting fossils for Cope, Osborn, von Zittel and others, has recently published his experiences in a delightful little volume (illustrated), under the title *The Life of a Fossil Hunter*, American Nature Series, Henry Holt and Co., New York. It is a most readable book, written with great enthusiasm.

[2] We retain the old spelling with the *e* as being nearer to the Greek, although Professor Cope writes it "Liodon."

PLATE XXXI.

RESTORATION OF TYLOSAURUS, A GREAT MARINE REPTILE OF CRETACEOUS AGE.
Length about 40 feet. (Drawn by J. Green).

second specimen, more complete in all respects, was discovered by Professor Cope's exploring party during an expedition from Fort Wallace, Kansas, in 1871. This specimen he has fully described and figured in Cope's *Tertiary Vertebrata*, 1875. It is a very instructive specimen, including fifty of the vertebræ from all parts of the vertebral column, a large part of the cranium, with teeth, as well as important limb-bones. These precious relics were excavated from a chalk "bluff," or high bank.

In considering the "Age of Reptiles," we cannot but marvel greatly at the diversity of forms assumed by the various orders of this class, their strange uncouth appearance, their assumption, in some cases, of characters only known at the present day among the mammals, their great abundance, and the perfect state in which their remains have been preserved in the stratified rocks of various parts of the world. And the reader may naturally ask, "How is it that so many types have disappeared altogether, leaving us out of a total of at least nine orders, only four, viz. those represented by crocodiles, lizards, snakes, and turtles?" To such a question we can only answer that the causes of the extinction of plants and animals in the past are not yet known. Climate, geographical conditions, food-supply, competition, with other causes, doubtless operated then as now; but if there is one clear lesson taught by the record of the rocks, it is this—that there has been at work from the earliest periods a Law of Progress, so that higher types, coming in at certain stages, have ousted the lower types, sometimes only partially, sometimes completely. But why the Dinosaurs, for instance, perished entirely, while the crocodiles survived to the present day, no one can yet explain. We can see no reason, however, why such problems as these should not be solved in the future by the co-operating labours of naturalists and geologists.

In the great onward and upward struggle for existence, higher types have supplanted lower ones; and, in accordance with this biological truth, we find that in the next era (known as the Tertiary or Cainozoic) the mammal held the field while the reptile took a subordinate place.

CHAPTER XI

FLYING DRAGONS

"Geology does better in reclothing dry bones and revealing lost creations than in tracing veins of lead or beds of iron."—RUSKIN.

THE great Ocean of Air was not uninhabited during the long ages of the Mesozoic era, when fishes swarmed in the seas, and reptiles, such as we have attempted to describe in the last five chapters, trod the earth, or swam across lakes and rivers. With such an exuberance of life in various forms, it would indeed have been strange if the atmosphere had only been tenanted by humble little insects like dragon-flies, locusts, or butterflies and moths, all of which we know were living then.

Now, the record of the rocks tells us that one great order of reptiles somehow acquired the power of flying, and flitted about as bats or flying-foxes do now. Since they were undoubtedly reptiles—in spite of certain resemblances to birds—we have ventured to call them "flying dragons," as others have done. The notion of a flying reptile may perhaps seem strange, or even impossible to some persons; but no one has a right to say such and such a thing "cannot be," or is "contrary to Nature," for the world is full of wonderful things such as we should have considered impossible had we not seen them with our eyes. Charles Kingsley, in his delightful fairy tale, *The Water-Babies*, makes some humorous remarks on that matter, which we may quote here. He says, "Did not learned men too hold, till within the last twenty-five years, that a flying dragon was an impossible monster? And do we not now know that there are hundreds of them found fossil up and down the world? People call them Pterodactyls; but that is only because they are ashamed to call them flying dragons, after denying so long that flying dragons could exist."

The illustrious Cuvier observes that it was not merely in magnitude that reptiles stood pre-eminent in ancient days, but they were distinguished by forms more varied and extraordinary than any that are now known to exist on the face of the earth. Among these extinct beings of ages incalculably remote, are the Pterodactyls,[1] or "wing-fingered" creatures, which had the power of flight, not by a membrane stretched over elongated fingers as in bats, nor by a wing without distinct or complete fingers, as in birds, but by a membrane supported chiefly by a greatly extended little finger, the other fingers being short and armed with claws.

The only reptile now existing which has any power of sustaining itself in the air is the little Draco volans, or "flying lizard," so called; but this can scarcely be regarded as a flying animal. Its ribs, however, are prolonged to such an extent that they support a broad expansion of the skin, so spread out from side to side as to perform the office of a parachute, thus enabling the creature to spring from tree to tree by means of extended leaps; and this it does with wonderful activity.

Many forms of Pterodactyl are known. Some were not larger than a sparrow; others about the size of a woodcock; yet others much larger, the largest of all having a spread of wing (or rather of the flying membranes) of twenty-five feet! It has been concluded that they could perch on trees, hang against perpendicular surfaces, such as the edge of a cliff, stand firmly on the ground,

[1] From the Greek—*pteron*, wing, and *dactylos*, finger.

and probably crawl on all fours with wings folded. It may be well at once to point out that the Pterodactyl had no *true* wings like those of a bird, but a thin membrane similar to that of a bat, only differently supported; so it must be understood that, when we use the word "wing," it is not in the scientific sense that we are using it, but in the popular sense, just as we might speak of the wing of a bat, although the bat has no true wing. Figs. 72, 73, 74, 75, and 77 will give the reader some idea of the various forms presented by the skeletons of Pterodactyls, or, as some authorities call them, Pterosaurians (winged lizards). Great differences of opinion have existed among palæontologists as to whether they are more reptilian than bird-like, or even mammalian.

More than a hundred years ago, in 1784, Collini, who was Director of the Elector-Palatine Museum at Mannheim, described a skeleton which he regarded as that of an unknown marine animal. It was a long-billed Pterodactyl from the famous lithographic stone of Solenhofen in Bavaria. The specimen was figured in the *Memoirs of the Palatine Academy*. Collini was able from this specimen to make out the head, neck, small tail, left leg, and two arms; but beyond that, he was at a loss. His conclusion was that the skeleton belonged neither to a bat nor to a bird, and he inquired whether it might not be an amphibian.

In 1809 this specimen came into Cuvier's hands, who at once perceived that it belonged to a reptile that could fly, and it was he who proposed the name Pterodactyl. Until the oracle at Paris was consulted, the greatest uncertainty prevailed, one naturalist regarding it as a bird, another as a bat. Cuvier, with his penetrating eye and patient investigation, combated these theories, supported though they were by weighty authorities. The principal key by means of which he solved the problem, and detected the saurian relationship of the Pterodactyl, seems to have been a certain bone belonging to the skull, known as the quadrate bone.

Pterodactylus macronyx, or, as it is now called, Dimorphodon macronyx (Fig. 72), was about the size of a raven. It was discovered in 1828 by the late Miss Mary Anning, the well-known collector of fossils from the Liassic rocks that form the cliffs along the coast of Dorsetshire, near Lyme Regis. This important specimen was figured and described by Dr. Buckland, in the *Transactions of the Geological Society*. He suggested the specific name macronyx on account of the great length of the claws.

This authority pointed out an unusual provision for giving

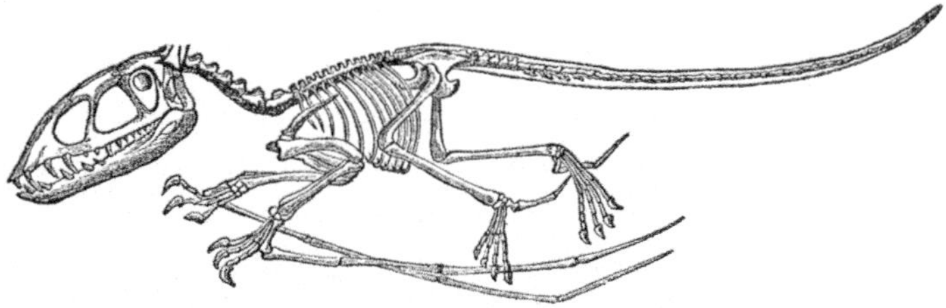

FIG. 72.—Skeleton of *Dimorphodon macronyx*. (After Owen.)

support and power of movement to the large head at the extremity of a rather long neck, namely, the occurrence of fine long tendons running parallel to the neck-vertebræ. This does not occur in any modern lizards, whose necks are short, and require no such aid to support the head. The restoration of the skeleton seen in the figure is by Professor Owen. It is probable that this Pterodactyl could walk on the ground with its wings folded, and perhaps it was also capable of perching on trees, by clinging on to their branches with its feet and toes. When the flying membrane was stretched out it must, on account of the long tail to which it was also attached, have presented a triangular shape, somewhat like a boy's kite.

Another genus, also from the lithographic slate of Bavaria, namely, Scaphognathus (so called on account of its large beak and jaws), had a very short tail, and its skeleton looks somewhat clumsy for a creature adapted to fly through the air (Fig. 73).

Pterodactylus spectabilis, from the same strata, also possessed a very short tail, but has a more elegant and bird-like skull. This pretty little flying dragon was only about as large as a

FIG. 73.—Skeleton of *Scaphognathus crassirostris*. $\frac{1}{3}$ natural size.

sparrow (see Fig. 74). Its neck is comparatively short, with but few joints. The long slender beak was probably sheathed in horn, and the skull in several ways approaches that of a bird. Since there are no teeth in the jaws, we may suppose that it devoured dragon-flies or other insects, such as we know were in existence during the period when the lithographic stone of Bavaria was being deposited. Those forms that were provided with teeth probably devoured such fishes as they could catch by swooping down upon the surface of the water.

Cuvier thought, from the magnitude of their eyes, that Pterodactyls were of nocturnal habits. "With flocks of such creatures flying in the air, and shoals of no less monstrous Ichthyosauri and Plesiosauri swarming in the ocean, and gigantic crocodiles and tortoises crawling on the shores of the primæval lakes and rivers

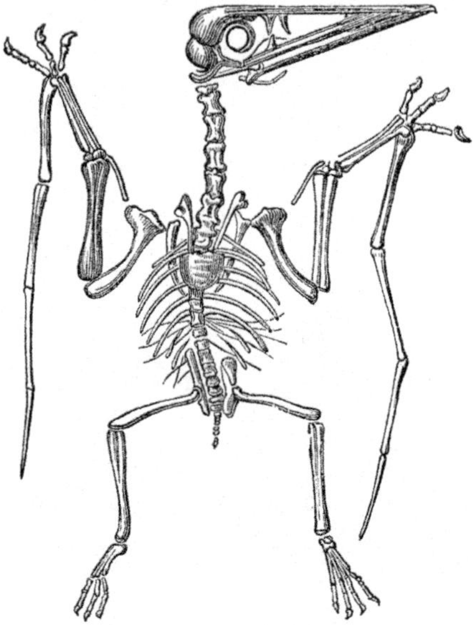

FIG. 74.—Skeleton of *Pterodactylus spectabilis.*

—air, sea, and land must have been strangely tenanted in these early periods of our infant world."[1]

It was thought at one time that Birds differed from Pterodactyls in the absence of teeth; but this only holds good for modern birds. If we go back to the Mesozoic age, we find that birds at that time did possess teeth. The oldest known bird, the Archæopteryx, had teeth in its jaws, and presents some very striking points of resemblance to reptiles (see p. 215). But if

[1] Buckland, *Bridgewater Treatise.*

GROUP OF SMALL FLYING DRAGONS, OR PTERODACTYLS.

Pterodactylus crassirostris. *Rhamphorhynchus phyllurus.* *Dimorphodon macronyx.*
Pterodactylus spectabilis.

PLATE XXXII.

we compare the skeleton of a Pterodactyl (such as the P. spectabilis, now under consideration) with that of a bird, we shall see in its fore limbs certain very obvious differences.

In the first place the arm or wing of a bird has only two, or three, bones, instead of four, or five, as here; secondly the outer finger, which corresponds with the human "little finger" is enormously elongated for the purpose of supporting the thin flying membrane. In two ways, however, the skeleton of a Pterodactyl does resemble that of a bird, for the bones are more or less pneumatic (according to von Zittel); and also the position of the skull at right angles to the axis bone (on the top of the neck vertebræ) is a bird-like arrangement. Again, natural casts of the brain cavity prove that the brain was decidedly avian, though smaller in proportion to the size of the skull. The orbits (for the eye) are very large and surrounded by a ring of small sclerotic plates (see Fig. 74).

In order to give rapid movement to their wings during flight, they had powerful muscles in the region of the chest. These were attached to a shield-like breast-bone provided with a keel —as in birds. But this bird-like feature is only a necessary provision to enable them to fly, and does not point to any relationship.

In the year 1873 was discovered, in the lithographic stone of Bavaria, at Eichstädt, a very beautiful new form of Pterodactyl. This was the Rhamphorhynchus phyllurus. The specimen is in a remarkable state of preservation; for the bones of the skeleton are nearly all in position, while those of both wings show very perfect impressions of the membranes attached to them. Its long tail supported another small leaf-like membrane, which was evidently used as a rudder in flight (see Fig. 75). The discovery of this valuable specimen attracted much attention at the time.

It was bought, by telegram, for Professor Marsh, and so secured for the Yale College Museum; but a cast may be seen at South Kensington.

Any one who looks carefully at the beautiful impressions of the wings of this specimen can see that they must have been produced by a thin smooth membrane, very similar to that of bats. When this elegant little creature was covered up by the fine soft mud that now forms the lithographic stone, its wings were partly folded, so that the membranes were more or less contracted into folds, like an umbrella only partly open. These

FIG. 75.—Skeleton of *Rhamphorhynchus phyllurus*, with delicate impressions of the flying membranes. (After Marsh.)

appear to have been attached all along the arm and to the end of the long finger. They then made a graceful curve backward to the hind foot, and probably were continued beyond the latter so as to join the tail. With its graceful pointed wings and long tail, this little flying saurian must have been a beautiful object, as it slowly mounted upwards from some cliff overlooking the Jurassic seas. (See Plate XXXII.)

Like those already described, it was provided with three short-clawed fingers, as well as the one which mainly supported its wing. Some of the Continental museums contain good collections of fossil Pterodactyls; but the largest collection in the world is that of Yale College, where Professor Marsh declares there are

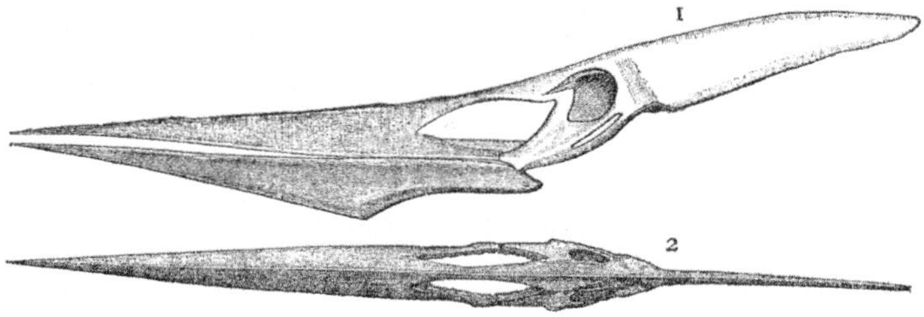

FIG. 76.—Skull of *Pteranodon*. 1. Side view. 2. Top view. (After Marsh.)

the remains of six hundred individuals from the American Cretaceous rocks alone!

Some of the fragmentary remains from our Cambridge Greensand formation indicate Pterodactyls of enormous size. Thus the neck-vertebræ of one species measure two inches in length, while portions of arm-bones are three inches broad. It is

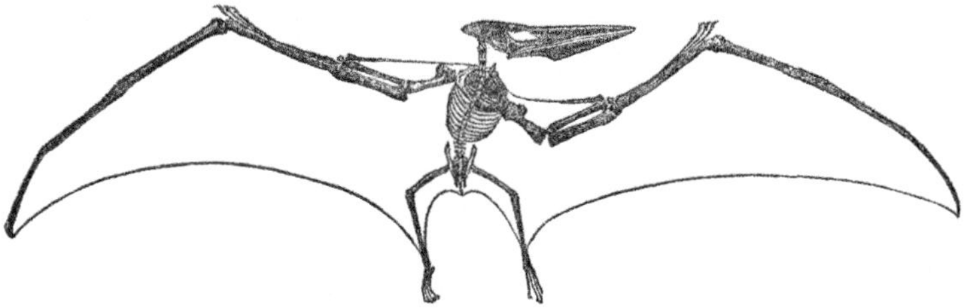

FIG. 77.—Skeleton of a toothless Flying Reptile (*Pteranodon occidentalis*), from the Upper Cretaceous of Kansas, U.S.A. Natural History Museum.

probable that the creatures to which these bones once belonged measured eighteen or twenty feet from tip to tip of the wings. Other also fragmentary remains from the chalk of Kent testify to the existence of Pterodactyls during that period fully equal in size.

But the largest Pterodactyls hail, like so many other big things, from America. Professor Marsh tells us of monsters

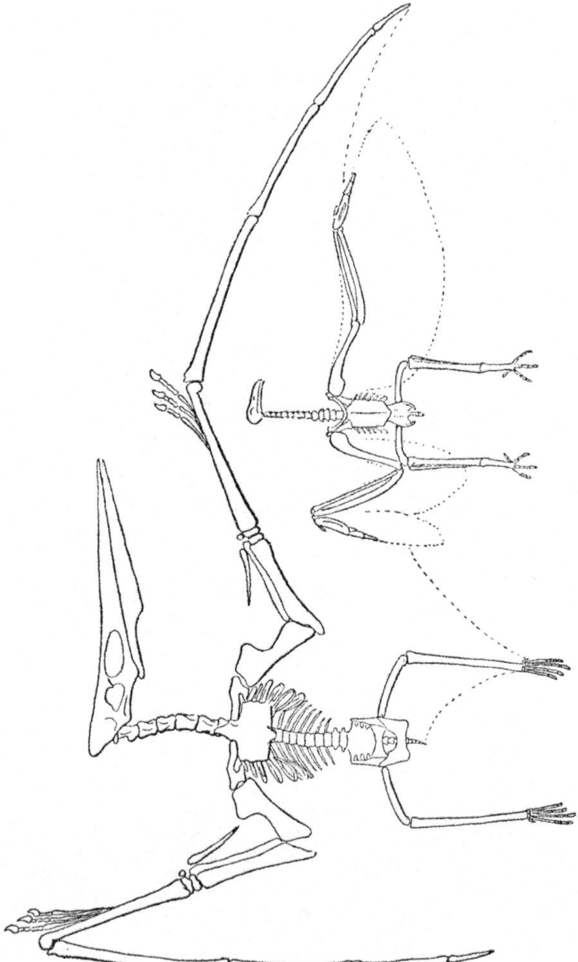

FIG. 78.—Skeleton of the great Pterodactyl *Pteranodon* compared with that of the Condor. Its wing was 9 feet long, and its skull nearly 4 feet long. Restoration by Dr. S. W. Williston. From the Smithsonian Report, 1901. The Condor's wing should be longer.

in his famous collection with a spread of wings of eighteen

PLATE XXXIII.

RESTORATION OF THE GIANT PTERODACTYL, PTERANODON, OF CRETACEOUS AGE.
Length from tip to tip of wings about 18 feet.
Smithsonian Report, 1901.

feet.[1] Plate XXXIII., shows a restoration of the largest known form, viz. Pteranodon, which has been described by Dr. S. W. Williston of the Chicago University. Its wing was nine feet long, and the skull, which is principally beak, was nearly four feet long, the beak being very long, pointed, and quite toothless, which suggests that this Pterodactyl dived after fishes.

Whether Pterodactyls were cold-blooded or warm-blooded is a question on which the authorities are not agreed. Professor Owen argued from the absence of feathers that they could not have been warm-blooded. But, in spite of this great authority, who has defended his opinion somewhat strongly, there are others who argue that the amount of work involved in sustaining a Pterodactyl in the air makes it highly probable that it was warm-blooded. Such, at least, is the view of the late Professor H. G. Seeley, who says of the Cambridge specimens: "That they lived exclusively upon land and in air is improbable, considering the circumstances under which their remains are found. It is likely that they haunted the sea-shores, and, while sometimes rowing themselves over the water with their powerful wings, used the wing-membranes, as the bat does, to enclose their prey and bring it to the mouth.

"The large Cambridge Pterodactyls probably pursued a more substantial prey than dragon-flies. Their teeth are well suited for fish, but probably fowl and small mammals, and even fruits, made a variety in their food. As lord of the cliff, it may be presumed to have taken toll of all animals that could be conquered with tooth and nail. From its brain it might be regarded as an intelligent animal. The jaws present indications of having been sheathed with a horny covering."

[1] Certain books still repeat the over-estimate of Marsh and state that the length from tip to tip was twenty-five feet. This is wrong.

Probably the large Pterodactyls of the Cretaceous period, soaring like albatrosses and giant petrels over the surface of the ocean, co-operated with the marine reptiles, such as Ichthyosaurs, Plesiosaurs, crocodiles, and others, as those sea-birds now do with the whales, porpoises, and dolphins, in reducing the excessive numbers of the teeming tribes of fishes, and in maintaining the balance of oceanic life.

With regard to the place of Pterodactyls in the animal kingdom, Professor Seeley placed them as a distinct sub-class, side by side with birds, and between mammals and reptiles, thus—

Mammalia.

Ornithosauria Aves.

Reptilia.

The name Ornithosauria (bird-lizards) is frequently used instead of the other name, because it expresses the idea of their being partly saurian and partly bird-like.

They flourished from the period of the Lias to that of the Chalk; and then, like so many other strange forms, seem to have suddenly disappeared.

CHAPTER XII

ANCIENT BIRDS

"Nor is the value of the doctrine of Evolution to the philosophic thinker diminished by the fact that it applies the same method to the living and the non-living world; and embraces in one stupendous analogy, the growth of a solar system from molecular chaos, the shaping of the earth from the nebulous cubhood of its youth, through innumerable changes and immeasurable ages, to its present form, and the development of a living being from the shapeless mass of protoplasm we term a germ."—The late PROFESSOR HUXLEY.

FOSSIL birds are but rarely met with in the stratified rocks; hence our knowledge of the bird life of former ages is comparatively slight. But this is only what might have been expected; for it must be remembered that birds are protected by their powers of flight from perishing in such ways as other animals frequently do. And even should they die on the water, their bodies are not likely to be submerged; for, being light and feathery, they do not sink, but continue floating until the body rots away, or is devoured by some creature, such as a hungry pike.

When did the bird make its first appearance, or *début*, on the earth? in other words, when did that primitive, but as yet unknown reptile from which the feathered tribe came, first take to itself feathers and assume both the habits and appearance of a bird? This is one of those interesting questions which remain to be solved by the labours of the palæontologist—or, more probably, of a generation of palæontologists. We have already alluded (see p. 126) to Professor Huxley's theory that birds are descended from Dinosaurs; but though there is much to be said in favour of the idea, we prefer, for our part, to wait and see what evidence may yet turn up on this subject, and, like the Irishman, to "prophesy *after* the event," *i.e.* when further discoveries have been made. Sir R. Owen never favoured the theory, and, for all palæontologists can tell, it may just as well be that both birds and pterodactyls (flying reptiles) are descended from a common stock; the one line choosing to fly by means of a thin membrane attached chiefly to a single long finger, while the others thought they could do quite as well—in fact better—by growing feathers on their arms and fingers. All great problems in Nature are solved slowly, by the patient accumulation of evidence; and the one above alluded to is no exception to the rule.

Palæontologists are not without evidence bearing on this subject; so perhaps we cannot do better than state briefly what that evidence is, in order to show how near, or how far, we are from a final answer to our question. Though no one can yet say what the very first bird-type was like, we can, at all events, describe the oldest known fossil bird. This is the famous Archæopteryx.[1]

Time was—and that within the memory of living geologists—when no fossil birds were known in rocks older than the Tertiary deposits; but the discovery of Archæopteryx has changed all that, and we now trace back the bird line to the middle of the great Secondary or Mesozoic Era. This bird was found in the Solenhofen limestone of Bavaria, which is supposed to represent the lower part of our English Kimmeridge clay. First, only the impression of a single feather was known, to which the late

[1] Greek—*Archios*, ancient; *pterux*, wing.

Professor H. von Meyer gave the name Archæopteryx lithographica (because this stone is much used by lithographers). Later on came the discovery of a magnificent and almost complete speci-

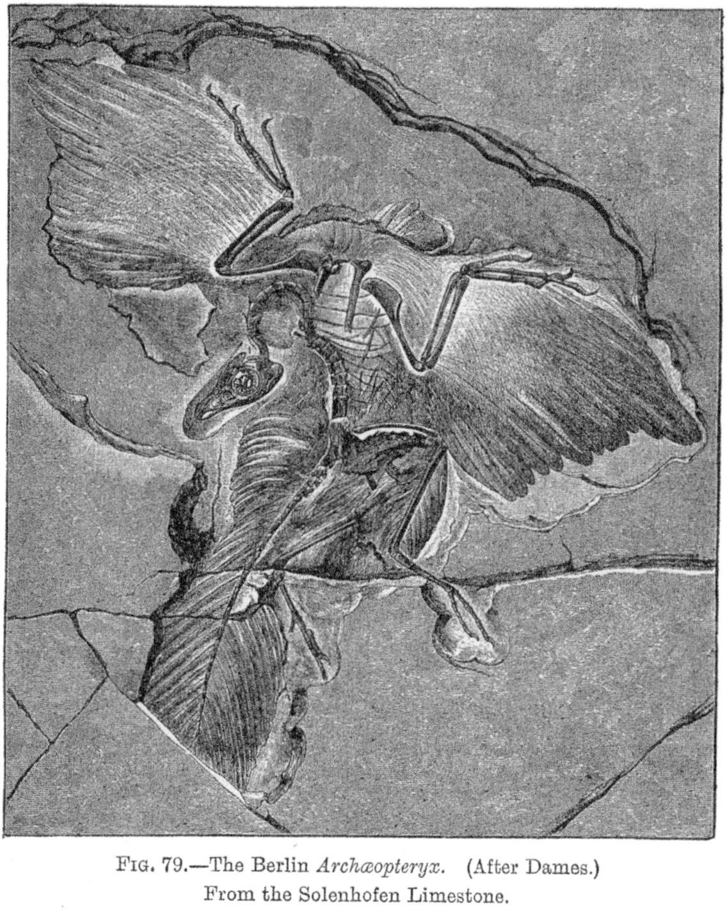

Fig. 79.—The Berlin *Archæopteryx*. (After Dames.)
From the Solenhofen Limestone.

men, with beautiful impressions of the feathers. This was named by Owen A. macrura, on account of its long tail. The specimen is to be seen in the Natural History Museum; it was only acquired for the nation by the strenuous efforts and liberality of the late Sir R. Owen (see preface). More recently a still finer specimen, with the head complete, has turned up, and this is now in the Berlin Museum (see Fig. 79). A very good idea of it may

FIG. 80.—Restoration of *Archæopteryx*, the oldest known bird. By W. P. Pycraft. Skull of same.

be gathered from the fine drawing presented by the late Professor Dames, which is to be seen at the Natural History Museum, also a cast. Some of those who first studied the creature's anatomy took it for a reptile; and they may well have been puzzled by its curiously "mixed" characters. It is certainly one of the most anomalous types of ancient life that has ever been discovered.

Palæontologists are now agreed that it *was* a bird; for it had feathers and claws, as birds have. But what can we say to a bird with teeth in its jaws, and with a long, lizard-like tail such as reptiles have now, except that this tail was provided with feathers attached in a very peculiar way, contrary to all bird's tails of the present day? A naturalist acquainted only with the avian life of to-day, would certainly say that it upsets nearly all his ideas of what a bird *ought* to be; but that only shows how useless it is (as we have previously pointed out, see pp. 105 and 199) to lay down rules for Nature. Its vertebræ are bi-concave, like those of fishes and some extinct Saurians. Another very reptilian feature is the presence of "sclerotic plates" in the eye. A pair of feathers sprang from each joint in the tail, which is quite a different arrangement to that in the tail of living birds. The leg-bone and foot are similar to that of modern perching birds, but then we have seen that some of the Dinosaurs, such as the little Compsognathus, had very bird-like feet. The wing shows three free digits or fingers. In form and position these three finger bones are just what may be seen in some young birds of to-day. It has been claimed by some that the fore limb of Archæopteryx is reptilian, but Mr. W. P. Pycraft has shown that it is more bird-like than some people thought, for it much resembles that of a young chick, and still more that of a primitive bird known as Opisthocomus. Another point, established by Marsh, is that the bones of the pelvis are separate—not united, as in modern birds. In the London specimen (where the skull is lost) there remains a cast of what is supposed (perhaps wrongly) to be the brain cavity. Were it not for the feathers, perhaps no one would at first have thought of calling it a bird. It combines the features, or characters, of birds and reptiles in a most remarkable way. In size it was about as large as a rook. Are we to call it a Pterodactyl with feathers? That would be to dub it a reptile; and since the leading authorities agree in calling it a bird, we must abide by their decision. Names, however, are sometimes misleading; the main fact we have to remember is, that in former ages some classes of animals which are now sharply marked off from each other were by no means so separated.

A good deal of discussion has arisen with regard to the true interpretation of the Berlin specimen, as far as the wing is concerned. The question is, how the feathers were attached; and there can be but little doubt, that in this respect the drawing from which Fig. 79 is taken is wrong. Look at the impressions of the long "primary" feathers, and you will see that at first they curve towards the three fingers, but then turn downwards and bend towards the two bones of the fore-arm. This is due to a mistake, and the first curve of the feathers, as seen in the left-hand corner, should have been continued. But in other respects Fig. 79 gives a fair idea of the Berlin specimen.

One cannot help wondering whether this very ancient bird could sing; but, although doubtless the gift of song was not in those Jurassic days so marvellously perfected as it is now, yet we would fain believe that Archæopteryx at least *tried* to sing. If, as seems probable, it lived among trees, one would think that some means of communication with its fellows was almost a necessity; besides, even lizards and some amphibians make a kind of music of which the song of birds seems to be a development.

The next evidences of former bird life are met with in the Cretaceous rocks, and here we find the Hesperornis. This bird was a gigantic diver, and its length from the point of the bill to

A GIGANTIC DIVER, HESPERORNIS REGALIS. FROM NORTH AMERICA. CRETACEOUS PERIOD.

PLATE XXXIV. Height 6 feet.

the end of the toes must have been between five and six feet (see Plate XXXIV.). Its habits are clearly indicated by the skeletons described by Professor Marsh in his splendid monograph on *The Extinct Toothed Birds of North America.* No living birds possess teeth in their jaws, so that the presence of such in Hesperornis

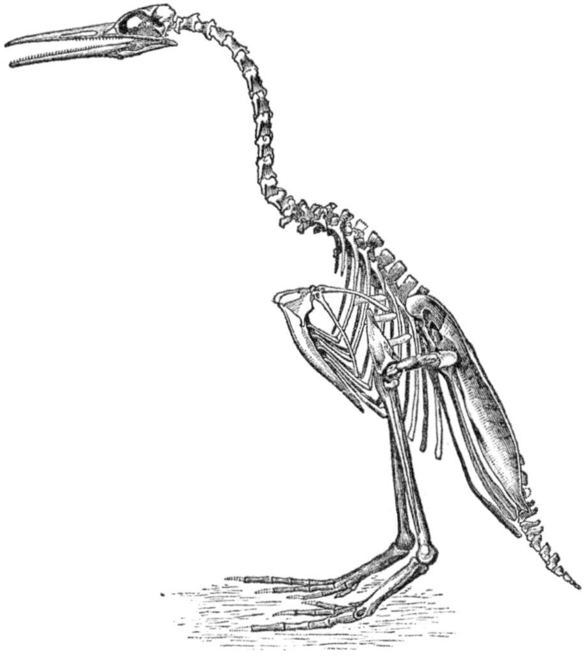

FIG. 81.—Skeleton of *Hesperornis regalis*, from Cretaceous strata, North America. (After Marsh.)

and other birds of the Cretaceous period at once separates them from those of the present day. It cannot be doubted that this antique diver was carnivorous; it probably devoured fishes. The teeth were set in a groove, and old ones were replaced by young ones growing up from inside the fang. The breast-bone (sternum) was entirely without a keel. The single thin wing-bone (humerus) indicates that its wings were "rudimentary," and quite useless either for flying or swimming (see Fig. 81). Modern penguins use their wings with great effect while swimming under water, but the Hesperornis was compensated for the want of wings by its broad tail, which was much expanded horizontally, and doubtless served as an organ of propulsion in diving. (The tail in our restoration is too small.)

The brain was diminutive, and very like that of a reptile. It is impossible, in the absence of feathers, to say exactly what appearance the skin of this bird presented, but in the restoration, Plate XXXIV., our artist has to some extent taken a modern diver as his model. The skeleton of this bird presents several interesting points of resemblance with reptiles; but it will not be necessary to enter into anatomical details here. The remains were discovered by Professor Marsh, in certain marine Cretaceous strata along the eastern slope of the Rocky Mountains, which have also yielded mosasauroid reptiles, Pliosaurs, and Pterodactyls. The valuable specimens on which Professor Marsh's descriptions are based were not procured without considerable trouble and hardship. His exploring parties in the plains of Kansas and Colorado suffered, at one time, from extreme heat, causing sunstroke and fever, by which the guides and explorers alike were weakened and discouraged; at other times, from extreme cold, to say nothing of the danger from hostile Indians. But fortunately, the latter enemies to science were kept in check by escorts of soldiers.

Hesperornis, as we have seen, was an admirable diver; while the long neck, with its capabilities of rapid flexure, and the long slender jaws armed with sharp recurved teeth, formed together a perfect instrument for the capture and retention of the most agile fish. As the lower jaws were united in front only by cartilage, as in the serpents, and had on each side a joint which admitted of some motion, the power of swallowing was doubtless equal to almost any emergency.

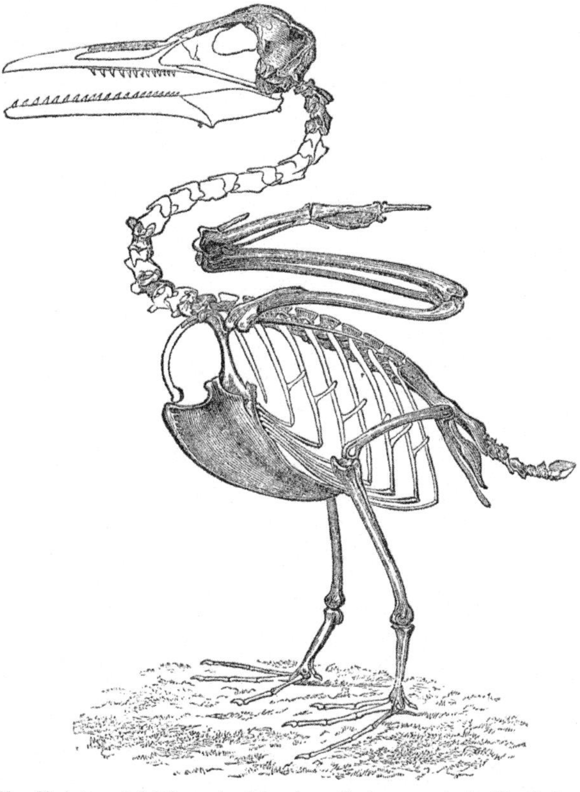

FIG. 82.—Skeleton of *Ichthyornis victor*, from Cretaceous strata, North America. (After Marsh.)

Another bird discovered by Professor Marsh in the same region, also with strangely blended characters, is the Ichthyornis (Fig. 82). Unlike the big diver above described, it had well-developed wings and a strongly keeled sternum for the attachment of muscles with which to work its wing. It was about the size of a rock-pigeon. The jaws were armed with teeth placed in distinct sockets, as in some extinct reptiles. The wing bones show that it possessed considerable powers of flight. Here we may note that the Cretaceous birds at present known (some twenty species or more) were apparently all aquatic forms, which, of course, are most likely to be preserved in marine deposits, while the Jurassic Archæopteryx was a land bird.

Remains of Cretaceous birds were first found in the Upper Greensand of Cambridge, and on these bones the genus Enaliornis has been founded. In its head and neck it resembled the divers.

Several portions of fossil birds have been discovered in the London Clay deposit of the Isle of Sheppey. One of these, the Dasornis, represented by a single skull, was as large as an ostrich, and probably closely related to that bird. Another, the Argillornis, rivalled the albatross in size. A third, the Odontopteryx (toothed bird), has a powerful serrated bill, well adapted for seizing its fishy prey. In the same case in the Natural History Museum may be seen casts of the limb-bones of a large bird, the Gastornis parisiensis, from Eocene strata near Paris; also casts of two leg-bones of another equally large bird, allied to the above, discovered in the Eocene strata near Croydon, viz. Gastornis Klaasseni. A restoration of the French bird's skeleton is shown in Fig. 83. The genus must have been as large as an ostrich, but more robust, with some signs of affinity with geese, as well as to ratite birds such as the ostrich.

Many other orders of birds are more or less represented by fossil remains from Tertiary strata, but in most cases not so perfectly as to warrant description here. For instance, the Pliocene strata of the Sivalik hills have yielded bones of

Plate XXXV.

THE DODO, DIDUS INEPTUS. ISLAND OF MAURITIUS.
Exterminated in the seventeenth century.

giant storks. Leaving these, we now pass on to say a few words about the Dodo, and some other birds that lived in historic times.

Every one has heard of that singular bird, now totally extinct, the Didus ineptus (or Dodo), which formerly inhabited the islands of Mauritius and Bourbon (see Plate XXXV.). The last record of its appearance dates from the year 1681. It was a large and heavy bird, bigger than a swan, and entirely unlike the pigeons, with which, however, it seems clearly to be allied. Its wings were so small as to be quite useless for flight. The legs were short and stout, with four toes on each foot, and the tail was extremely short, carrying a tuft of soft plumes. The beak was decidedly hooked, as in birds of prey.

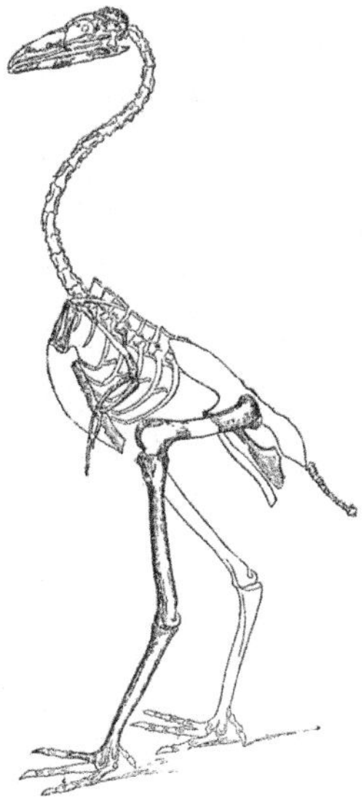

FIG. 83.—Restored skeleton of a large bird, *Gastornis Edwardsii*, from Eocene strata near Paris and Rheims. (After M. le Docteur Lemoine.) Height over 7 feet.

Visitors to the Natural History Museum will see, in the same case with the remains of the gigantic Æpyornis from Madagascar, some of the bones of this remarkable bird, together with a portrait from an old painting. The Oxford Museum once possessed a complete stuffed specimen, but now, alas! only the head and one foot remain.

In the reports of numerous voyagers who visited the islands of Mauritius, Bourbon, and Rodriguez, from the end of the fifteenth century to the middle of the seventeenth, we have many accounts of the appearance and habits of this bird, evidently sketched from life. Some of the descriptions are very quaint; as, for example, the graphic sketch of old Sir Thomas Herbert, who saw the bird in his travels in the year 1634. In a second island of the Mascarene group, viz. that known as Rodriguez, early explorers found the Solitaire (Pezophaps solitaria), a near relative of the Dodo, with longer legs and neck.[1]

From Patagonian strata (perhaps of Miocene age) we have a skull and lower jaw of a huge carinate bird, the Phororhachos (see Fig. 84). The skull is twenty-three inches long and seven

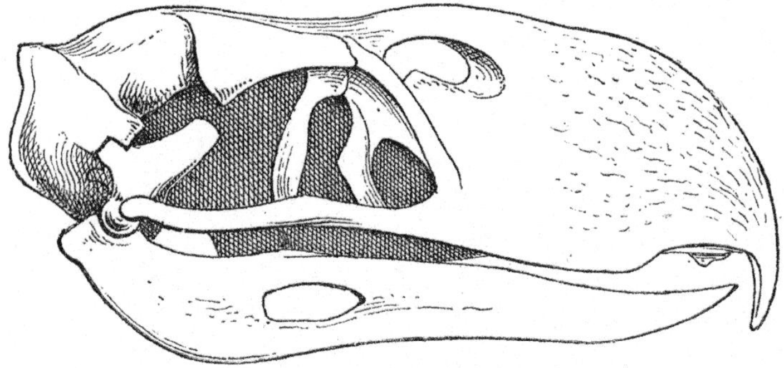

FIG. 84.—Restored skull and lower jaw of *Phororhachos longissimus*, from the Santa Cruz Formation of Patagonia; $\frac{1}{6}$ natural size.

inches deep. These measurements far exceed those of any living bird. No quite complete skull is known, but enough to justify the construction of a model skull and lower jaw as seen in our figure. Possibly this feathered giant was carnivorous, as his pointed beak rather suggests. He probably stood some seven feet high, but his relationships are doubtful at present. In Plate XXXVI. we have a restoration by Mr. Charles Knight showing very small wings which would be useless for flight.

Of all the monsters that ever lived on the face of the earth, the

[1] See the writer's *Creatures of Other Days*, p. 166.

PHORORHACOS, A PATAGONIAN GIANT OF THE MIOCENE PERIOD.
From a drawing by Charles R. Knight.
From "Animals of the Past," by Prof. F. A. Lucas, by permission of Messrs. Doubleday, Page and Co., New York.

PLATE XXXVI.

giant birds were perhaps the most grotesque. An emu or a cassowary of the present day looks sufficiently strange by the side or ordinary birds; but "running birds" much larger than these flourished not so very long ago in New Zealand and Madagascar, and must at one time have inhabited areas now sunk below the ocean waves.

The history of the discovery of these remarkable and truly gigantic birds in New Zealand, and the famous researches of Professor Owen, by which their structures have been made known, must now engage our attention.

In the year 1839 Professor Owen exhibited, at a meeting of the Zoological Society, part of a thigh-bone, or femur, 6 inches in length, and 5½ inches in its smallest circumference, with both extremities broken off. This bone of an unknown struthious bird was placed in his hands for examination, by Mr. Rule, with the statement that it was found in New Zealand, where the natives have a tradition that it belonged to a bird now extinct, to which they give the name Moa. Similar bones, it was said, were found buried on the banks of the rivers.

A minute description of this bone was given by the professor, who pointed out the peculiar interest of this discovery on account of the remarkable character of the existing fauna of New Zealand, which still includes one of the most extraordinary birds of the struthious order ("running birds"), viz. the Apteryx, and also because of the close analogy which the event indicated by the present relic offers to the extinction of the Dodo in the island of Mauritius. On the strength of this one fragment he ventured to assert that there once lived in New Zealand a bird as large as the ostrich, and of the same order. This conclusion, which some naturalists strongly opposed at the time, was more than confirmed by subsequent discoveries, which he anticipated; and, as we shall see, his estimate was a most moderate one, for the extinct bird turned out to be considerably larger than the ostrich.

Later on he received from a friend in New Zealand news of the discovery of more bones. In 1843 a collection of bones of large birds was sent to Dr. Buckland, Dean of Westminster, by the Rev. William Williams, a zealous and successful Church missionary, long resident in New Zealand.

The leg-bones sent to London greatly exceeded in bulk those of the largest horse. The leg-bone of a tall man is about 1 foot 4 inches in length, and the thigh of O'Brien, the Irish giant, whose skeleton, eight feet high, is mounted in the Museum of the Royal College of Surgeons, is not quite two feet. But some of the leg-bones (tibiæ) of Moas measure as much as thirty-nine inches.

In 1846 and 1847 Mr. Walter Mantell, eldest son of Dr. Mantell, who had resided several years in New Zealand, explored every known locality within his reach in the North Island. He also went into the interior of the country and lived among the natives for the purpose of collecting specimens, and of ascertaining whether any of these gigantic birds were still in existence; resolving, if there appeared to be the least chance of success, to penetrate into the unfrequented regions, and obtain a live Moa. The information gathered from the natives offered no encouragement to follow up the pursuit, but tended to confirm the idea that this race of colossal bipeds was extinct. He succeeded, however, in obtaining a most interesting collection of the bones of Moa, belonging to birds of various species and genera, differing considerably in size. This collection was purchased by the trustees of the British Museum for £200. Another collection was made by Mr. Percy Earle from a submerged swamp, visible only at low water, situated on the south-eastern shore of the Middle Island. This collection also was purchased by the trustees for the sum of £130.

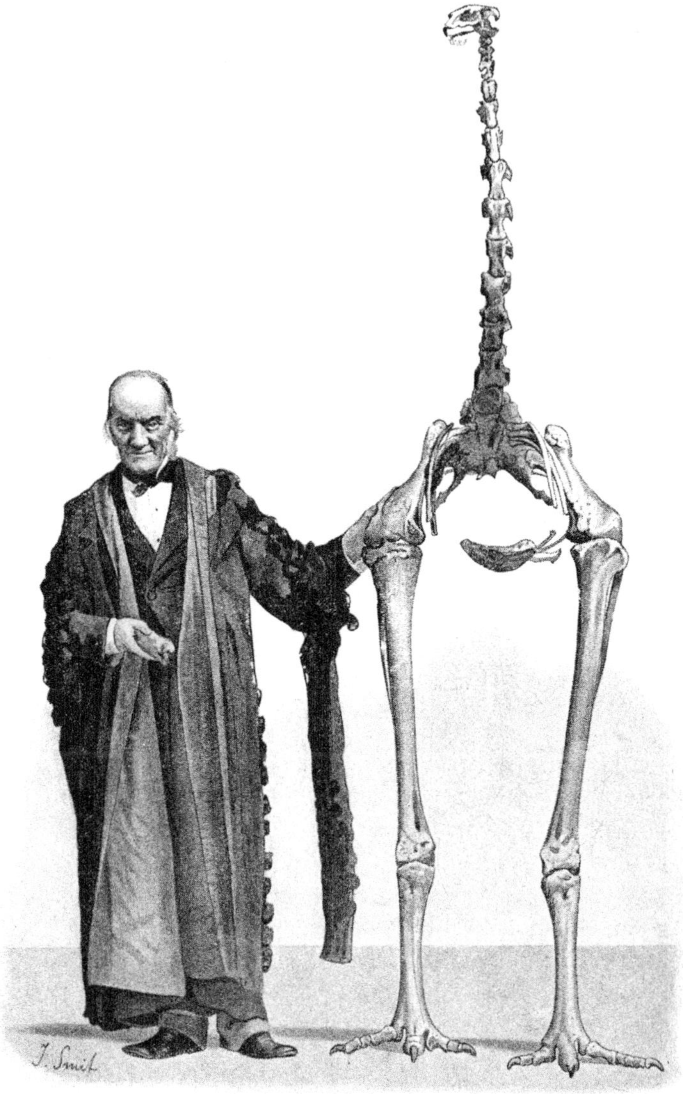

THE LATE SIR RICHARD OWEN AND A SKELETON OF MOA, DINORNIS MAXIMUS.

PLATE XXXVII. *(From a photograph.)*

The total number of species of Moa once inhabiting New Zealand was probably at least fifteen, and, judging from the enormous accumulations of their bones found in some districts, they must have been extremely common, and probably went about in flocks. "Birds of a feather *flock* together."

Not only was the number of individuals very large, but they belonged (according to Mr. F. W. Hutton) to no less than seven genera, containing twenty-five different species, a remarkable fact which is unparalleled in any other part of the world. The species described by Professor Owen in his great work,[1] vary in size from 3 feet to 12 or even 14 feet in height, and differ greatly in their forms, some being tall and slender, and probably swift-footed like the ostrich, whilst others were short and had stout limbs, such as Dinornis elephantopus (Fig. 85), which was undoubtedly a bird of great strength, but very heavy-footed. Dinornis crassus also had stout limbs.

The Natural History Museum at South Kensington contains a valuable collection of remains of Moa. These skeletons may be seen in Gallery No. 2. In D. giganteus the leg-bone (see Fig. 85) attains the enormous length of three feet, and in an allied species it is even thirty-nine inches! The next bone below (cannon bone) is sometimes more than half the length of the leg-bone (tibia). (See Plate XXXVII.)

A skeleton in one of the glass cases has a height of about 10½ feet, and it is concluded that the largest birds did not stand less than 12 feet, and possibly were 14 feet high!

Dinornis parvus (the dwarf Moa) was only three feet high.

[1] Memoir on *The Extinct Wingless Birds of New Zealand*. London, 1878. The beautiful drawing by Mr. Smit (Plate XXXVII.) is from a photograph in this valuable work representing the late Sir Richard Owen standing in academic robes by the side of a specimen of the skeleton of the great Dinornis maximus.

In 1882 was obtained, from a cave in Otago, the head, neck, two legs, and feet of a Moa (D. didinus), having the skin, still preserved in a dried state, covering the bones, and some few feathers of a reddish hue still attached to the leg. The ring of the windpipe may be seen *in situ*, the sclerotic plates of the

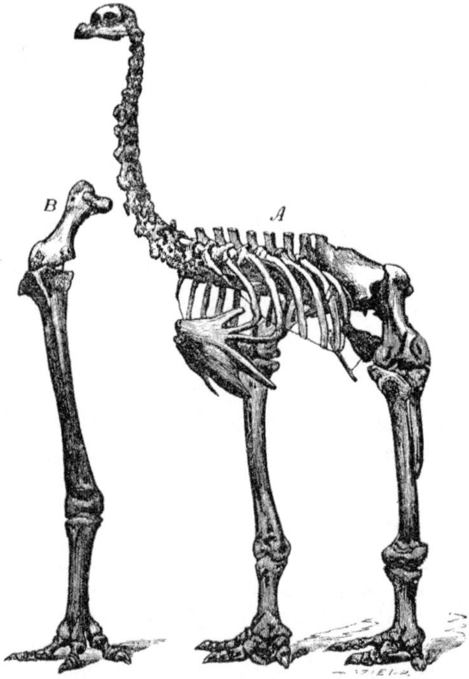

Fig. 85.—*A*. Skeleton of the Elephant-footed Moa, *Dinornis elephantopus*, from New Zealand. *B*. Leg-bones of *Dinornis giganteus*, representing a bird over 12 feet high.

eye, and the sheaths of the claws. One foot also shows the hind claw still attached.

From traditions and other circumstances it is supposed that the present natives of New Zealand came there not more than about

PLATE XXXVIII.

MOAS FROM NEW ZEALAND.

Dinornis giganteus.
Height 12 feet.

D. elephantopus.
A small species.

six hundred years ago, and there is reason to believe that the ancient Maoris, when they landed, feasted on Moa as long as any remained. Their extermination *probably* only dates back to about the period at which the islands were thrice visited by Captain Cook, 1769–1778. The Moa is mixed up with their songs and stories, and they even have a tradition of caravans being attacked by them. Still, some people believe that they were killed off by the race which inhabited New Zealand before the Maoris came. But they must have been there up to a time not far removed from the present. It is even said that the "runs" made by them were visible on the sides of the hills up to a few years ago; and possibly they may still be visible. The charred bones and egg-shells have been found mixed with charcoal where the native ovens were formerly made, and their eggs are said to have been found in Maori graves. Mr. Hutton considers that in the North Island they were exterminated three or four centuries ago, while in the South Island they may have lingered a century longer. (See Plate XXXVIII.)

The nearest ally of the Moa is the small Apteryx, or Kiwi, of New Zealand, specimens of which may be seen at the Natural History Museum, at the end of the long gallery devoted to living birds. This bird, however, has a long pointed bill for probing in the soft mud for worms, whereas the bill of the Moa was short like that of an ostrich.

Another difference between the two is that, while the Kiwi still retains the rudiments of wing-bones, the Moa had hardly a vestige of such.

In Australia the remains have been found of a bird probably related to the Cassowaries, but at present imperfectly known. To this type of struthious, or running bird, the name Dromornis has been given.

Now, it is a remarkable fact that remains of another giant bird and its eggs have been found on the opposite side of the great Indian Ocean, namely, in the island of Madagascar, the existence of which was first revealed by its eggs, found sunk in the swamps, but of which some imperfect bones were afterwards discovered. One of these eggs was so enormous that its diameter was nearly fourteen inches, and was reckoned to be as big as three ostrich eggs, or 148 hen's eggs! This means a cubic content of more than two gallons! The natives search for the eggs by probing in the soft mud of the swamps with long iron rods. A large specimen of an egg of this bird will fetch a good price. What the dimensions of Æpyornis were it is impossible to say, and it would be unsafe to venture a calculation from the size of the egg.[1] The reader who wishes to see some of the remains of this huge bird may be referred to the Natural History Museum. In Wall case No. II., Gallery 2, may be seen a tibia and plaster casts of other bones; also two entire eggs, many broken pieces, and one plaster cast of an egg found in certain surface deposits in Madagascar. In the same case may be seen bones of the Dodo from the Isle of Mauritius.

It will thus be seen that we have three distinct groups of giant land birds—the Moas, the Dromornis, and the Æpyornis,—occupying areas at present widely separated by the ocean.

This raises the difficult but very interesting question, how they got there; and the same applies to their living ancestors. The ostrich proper, Struthio camelus, inhabits Africa and Arabia; but there is evidence from history to show that it formerly existed in Beluchistan and Central Asia. And, going still further

[1] From the size of a femur and tibia of Æpyornis preserved in the Paris Museum, it could not have been less in stature than the Dinornis maximus of New Zealand.

back, the geological record informs us that, in the Pliocene period, they inhabited what is now Northern India. In Australia we have the Cassowary (Casuarius) and the Emeu (Dromaius); in New Zealand, the Apteryx (or Kiwi). Now, as none of these birds can either fly or swim, it is impossible that they could have reached these regions separated as they now are; and it is hardly likely that they arose spontaneously in each district from totally different ancestors. But the doctrine of evolution affords a key to the problem, and tells us that they all sprang from a common ancestor, of the struthious type (probably inhabiting the great northern continental area), and gradually migrated south along land areas now submerged. In this way we get some idea of the vast changes that have taken place in the geography of the world during later geological periods. Perhaps they were compelled to move south, until they reached abodes free from carnivorous enemies. Having done so, they evidently flourished abundantly, especially in New Zealand, where there are so few mammals, only those recently introduced by man.

In North America Professor Cope has reported a large wingless fossil bird from the Eocene strata of New Mexico. In England we have two such—namely, the Dasornis, from the London Clay of Sheppey (Eocene period), and the Gastornis, from the Woolwich beds near Croydon, and from Paris (also Eocene).

Lake Callabonna in Australia (see p. 310) has yielded the bones of some huge ratite birds, *e.g.* Genyornis, with very slender toes. It lived with the Diprotodon. In South America the living Rheas were preceded by such large birds as Brontornis, of which casts of limb-bones are to be seen in the Natural History Museum. The famous Sivalik strata of India have yielded remains of an ostrich, Struthio asiaticus, and these may be seen in the above Museum, with the skeleton of a modern ostrich, among the fossil birds. The Fayum of Egypt (see p. 262) has also yielded a portion of a limb-bone of some large struthious bird.

It will thus be seen that big struthious birds have a long history, going far back into the Tertiary era, and that they once had a much wider geographical range than they have now. Doubtless, future discoveries will tend to fill up the gaps between all these various types, both living and extinct, and to connect them together in one chain of evolution. Possibly the bird evolved from an arboreal reptile.

CHAPTER XIII

SOME EARLY MAMMALS AND THE STORY OF THE HORSE

"Geology, in the magnitude and sublimity of the objects of which it treats, ranks next to astronomy in the scale of the sciences."—Sir JOHN F. W. HERSCHEL.

IT has often been said that the Primary era was an "Age of Fishes," the Secondary era an "Age of Reptiles," and the Tertiary an "Age of Mammals." There is, however, a danger lest beginners should be deceived by broad statements of this kind, which must not be pressed too far. All that is meant in this case is, that, first of all, fishes were the dominant type; then reptiles, and then mammals.

Fishes abounded in the waters of the Primary or Palæozoic era, while air-breathing amphibians appeared towards its close; reptiles flourished vigorously all through Secondary or Mesozoic times: and now we are about to show that a marvellous outburst of mammalian life appears to have taken place very early in the Tertiary or Cainozoic era. All this is in accordance with the "Law of Progress" throughout past times, and strongly confirms the theory of Evolution (see pp. 84, 106, 125, and 264).

The student of geological history soon discovers that mammals of some kind, or kinds, *did* exist throughout the Secondary era—although, judging from their imperfect remains, they must have been of a low type.[1] But the great advance in our knowledge of

[1] Perhaps some of the jaw-bones from Triassic and Jurassic strata represent creatures as low down in the scale as the Monotremes, represented at the present day by the Duck-bill Platypus and the Echidna of Australia.

the world's extinct races which has taken place during the last quarter of a century ought to make us careful with regard to broad statements of the above kind, which may require to be modified by future discoveries.

Now, the tendency of all the later results of Palæontology is to show that some of the higher types of life appeared on the earth a good deal earlier than was formerly supposed. Thus the discovery of the jaw-bones of small mammals (possibly marsupials) in the Purbeck strata, in 1854, and afterwards in the Stonesfield slate (both of Jurassic age), came as a surprise to most geologists. Again, it is not long since birds were believed to have come into existence only in Tertiary times: *now* we have the Jurassic bird Archæopteryx (see p. 212), and Professor Huxley has recorded his opinion that some of us may live to see a fossil mammal of the remote Silurian age! It is, therefore, only prudent for geologists to be on their guard against assuming that *all* the Secondary mammals were of low types and few in numbers, as might be inferred from the phrase "Age of Mammals." That mammals, however, developed vigorously in many directions during the Tertiary era cannot be doubted. So, far, then the expression conveys important truth; but in science, as in other things, the "unexpected" often happens, so it is better to be prepared for surprises.

It would be impossible within the limits of a single chapter to give an account of all the leading orders of Tertiary mammals. We must therefore limit our present remarks to two or three, or more, reserving others for future chapters.

The discovery, in the early part of the present century, of the rich treasures embedded in the Tertiary strata of the Paris basin, and the consummate skill with which they were interpreted and restored by the immortal Cuvier, gave a very great impulse to the study of Geology. His restorations became patterns for others, like Owen, Huxley, Marsh, Cope, Gaudry, and many more, who have worked on the lines he laid down. The mammalian remains brought to Cuvier by numerous collectors were very imperfect and fragmentary—detached bones and teeth, with occasionally some portion of a skeleton. The success with which he put them into order, and built up therefrom the long-lost types of Eocene days, was due largely to his wonderful knowledge of living animals, but partly also to his *Law of Correlation* (see p. 7).

One of Cuvier's triumphs was the restoration of the Palæo-

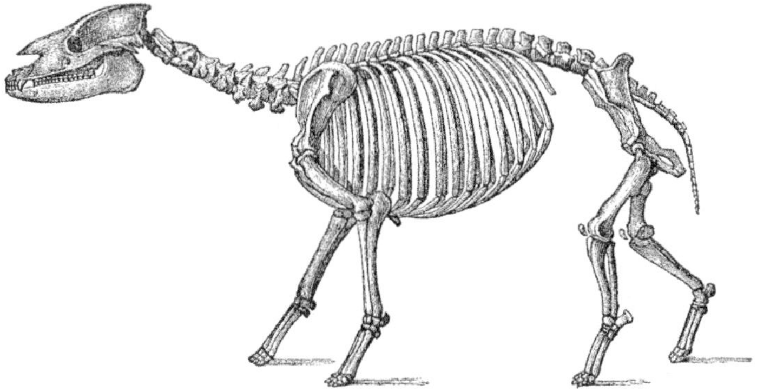

FIG. 86.—Skeleton of a tapir-like animal, *Palæotherium magnum*, from Eocene strata, near Paris. (After Gaudry.)

therium,[1] a tapir-like animal, from fragmentary remains found in the strata of the Paris basin, chiefly at Montmartre; his conclusions being afterwards verified by the discovery of a nearly complete skeleton (see Fig. 86). The molar teeth of this animal somewhat resembled those of the rhinoceros. The skull shows that it had a short proboscis. The toes, of which there were three on each foot, ended in small hoofs, the middle toes being

[1] Greek—*palaios*, ancient; *therion*, wild beast.

the largest. Its eye was small, and the head rather large. Several species have been determined, varying from the size of a sheep (P. curtum) to that of a horse (P. magnum), and it had much in common with both the horse and the rhinoceros. P. medium was rather smaller than an American Tapir: P. minus was a small and elegant species, of which the fresh-water Eocene beds of the Isle of Wight have yielded remains.

A complete specimen, discovered in 1874, with the outline of the body indicated in the rock, was of a slenderer build, and with a longer neck than in Cuvier's restoration, on which our artist's drawing in Plate XXXIX. is based; but the earlier discoveries do not agree with this, and the reader can easily see for himself that the skeleton shown in Fig. 86 could not have had a longer neck without at the same time having longer legs, and probably thinner bones. We have therefore adhered to the now familiar outline of the Palæothere as restored by Cuvier and since copied into almost every book on Geology. Doubtless there were many species existing at the time, and this later discovery probably represents one of the slenderer sort.

Such then, in all probability, was the Palæothere, a creature which lived in herds in the valleys of the plateau surrounding the ancient lake-basins of Orleans and Argenton; in the department of Gironde; in the Isle of Wight: and in various parts of Europe.

Its contemporary, the Anoplotherium,[1] so called from its apparently defenceless state, is represented in the same Plate. This animal was of a lighter and more elegant form, and its limbs ended each in two digits, only terminating in hoofs. As in the Palæothere, the jaws contained forty-four teeth, but there was no interval in the series. There are suggestions in its frame-

[1] Greek—*alpha*, privative; *hopla*, arms; *therion*, beast.

TAPIR-LIKE ANIMALS. EOCENE PERIOD. EUROPE.

Xiphodon. *Anoplotherium.* *Palæotherium.*

PLATE XXXIX.

work of the modern ruminant type;[1] thus it resembled the ox in having an equally divided hoof, or rather two hoofs. Fig. 87, in which the skeleton is seen, shows that it possessed a long tail. In size it was about equal to a fallow deer, and the long tail has been supposed, perhaps erroneously, to indicate aquatic habits. Small and more delicate species have been taken to represent distinct genera.

"Judging from its habits of swimming and diving," said

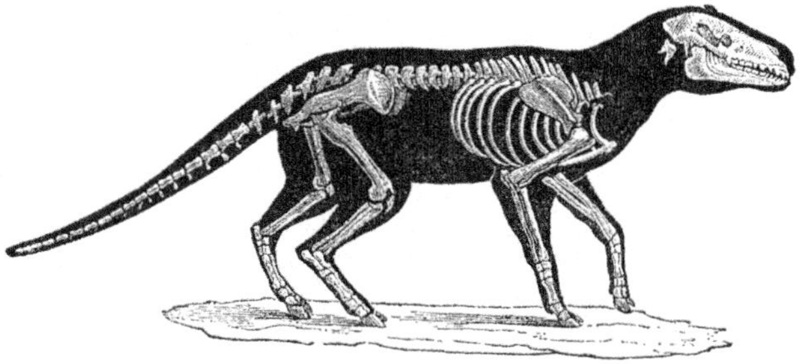

FIG. 87.—Skeleton with restored outline of *Anoplotherium commune*, from Eocene strata. (After Cuvier.)

Cuvier, "the Anoplotherium would have the hair smooth like the otter; perhaps its skin was even half naked. It is not likely, either, that it had long ears, which would be inconvenient in its aquatic kind of life; and I am inclined to think that, in this respect, it resembled the hippopotamus and other

[1] Cuvier divided all the hoofed animals (ungulates) into two orders, pachyderms and ruminants. The former is a heterogeneous order, and has since been abandoned; but the ruminants have been regarded as one of the most distinct of mammalian orders, for they are separated from all other animals by having horns and hoofs in pairs, the absence of upper front teeth, complex stomachs, and the habit of ruminating or "chewing the cud." Professor Owen showed that ungulates should be classified by the structure of their feet. He therefore divided them into odd-toed (perissodactylate) and even-toed (artiodactylate), and he placed elephants in a separate order—the Proboscidia.

quadrupeds which frequent the water much." But was it really aquatic?

Xiphodon,[1] so named by Cuvier on account of the shape of its teeth, was a small and delicate animal, long, with slenderer limbs than the Anoplotherium; its feet were provided with two toes, and the tail was short (see restoration in Plate XXXIX.). X. gracilis was obtained from the lignites of Débruge, near Apt. It was some three feet high, and of about the size of a chamois, but lighter in form, and with a smaller head. Cuvier says of this creature, "Its course was not embarrassed by a long tail; but, like all active herbivorous animals, it was probably timid, and with large and very mobile ears, like those of the stag, announcing the slightest approach of danger. Neither is there any doubt that its body was covered with smooth hair, and consequently we only require to know its colour in order to paint it as it formerly existed in this country, where it has been dug up after so many ages." Instead of resorting to rivers and lakes, this graceful little creature probably kept to the dry land and fed upon aromatic herbs.

Discoveries, chiefly made since Cuvier's day, have shown that these ancient herbivorous mammals had carnivorous enemies, which doubtless kept down their numbers—not lions and tigers, but certain lower and less "specialised" creatures, from some of which the latter are descended.

The nearest living ally of our Palæothere, that once inhabited Europe in great numbers, must be sought for a long way from home; and it is highly instructive to observe that the further we wander away from the present down "the corridors of time," the further we must travel, geographically, to find creatures at all matching those that lived in early ages of the world's history.

[1] Greek—*xiphos*, sword; *odous, odontos*, tooth.

Thus, the Tapir from Sumatra, or Central and South America, is the nearest living relative of the Palæothere; while, if we go back still further—into Jurassic times,—we must fetch marsupials from Australia to match some of the little mammals whose jawbones are found in some of the deposits of that age. To the geologist and naturalist, South America and Australia are countries which have, as it were, lagged behind; for their faunas are not "up to date," as the saying is; and thereby hangs a long story—into which we must not be led now,—and one which shows how closely the sciences of Geology, Natural History, and Geography are interwoven.

As already stated, Cuvier's great results in restoring Eocene animals were due, in a large measure, to the use he made of his principle of "correlation;" of which he writes: "I doubt whether I should ever have divined, if observation had not taught me, that the ruminant hoofed beasts should all have the cloven foot, and be the only beasts with horns on the frontal bone."[1] Again, it is found—though no one knows why—that only those hoofed animals which have their hoofs in one or two pairs, have horns in one or two pairs on the frontal bones: whilst those with three hoofs, if they have horns at all, have either one or two placed one behind the other (example, the Indian rhinoceros with one horn and the African two-horned rhinoceros). There must be secret reasons for these curious facts—who shall win immortality by discovering them? Other relationships, such as that between teeth and hoofs, might be given; but the above example is sufficient for our present purpose (see p. 7). In dealing with this subject, however, it should be mentioned that Cuvier's law is not infallible, and does not always apply to some of the ancient and generalised types discovered since his day.

[1] *Ossemens Fossiles*, tom. i. p. 184. (1834.)

The great theory of Evolution was first dimly suggested by Greek philosophers, such as Anaximander (B.C. 610), who may have derived the idea from Egyptian, Babylonian, or Hindu sources; then revived, in a more scientific form, by Lamarck last century. Fifty-two years ago it was placed on a truly scientific basis by the illustrious Charles Darwin, and is now generally accepted by naturalists and palæontologists. Indeed, it is hard to be a palæontologist in these days without being also an evolutionist—so abundant is the evidence derived from a study of extinct animals. Year by year the evidence is accumulating, and many workers in various parts of the world are discovering long-lost types which appear to link together some of the branches of the great Tree of Life. Marsh and Cope in America; Owen, Huxley, and others in England; have all been directing our ideas in the same course. At Pikermi in Attica; in the far Western States in America; at Sansans, Allier, Lebéron and other localities in France, such great and important additions have been made of late to our knowledge, that it is now possible to make out certain lines of evolution in the Mammalia since their first important outburst at the beginning of the Tertiary period.

We now propose to trace, in the light of recent discoveries, the history of one important living group, as represented by the horse.

The series of fossil horses now known is so complete, that hardly a single important gap is left between the original five-toed ancestor, and the horse of to-day, with only one toe to each foot! Here, then, we have the most perfect evidence of the evolution of an animal from distant ages in the earth's history that has ever been presented to the world! Professor Cope's researches on the history of camels as illustrated from fossil remains have brought to light another line of evolution equally interesting if not quite so complete. According to the latter palæontologist, the very earliest ancestor of all the hoofed animals (and therefore of the horse) was the Eocene Phenacodus, with five toes. But this conclusion is not generally accepted now.

We pass on to consider the true fossil horses, as worked out by Professors Huxley and Marsh, and others. The modern horse, and that which was known to man in the days before history was written, we may regard as a product of the latest geological period—the Pleistocene. The development of the horse from a primitive five-toed ancestor seems to have taken place along two separate lines, one in Europe and one in America. The latter is the most complete; for it so happens in that country the physical conditions which prevailed throughout nearly the whole of the Tertiary Era were singularly favourable to the preservation of the skeletons of those creatures which lived on land. The various members of the horse tribe that roamed over North America all through these long ages of the past were specially numerous in what is now the Rocky Mountain region, and their remains are sealed up in the strata of these regions.

Here in Wyoming and Utah is found one of the oldest direct ancestors of the horse, the Orohippus of Marsh. During the middle Tertiary, or Miocene period, two other lakes existed on either side of the great Eocene basin. The largest of these, to the east of the Rocky Mountains, extended over portions of what are now Dakota, Nebraska, and Colorado. The clays deposited in this lake form the "Bad Lands" of that region, so well known for their fossil treasures. The other Miocene lake was west of the Blue Mountains, where eastern Oregon now is, but, having since been overflowed by a vast sheet of basalt, its thickness is unknown. In this basin the Miohippus of Marsh first makes its appearance.

During the later Tertiary, or Pliocene period, a vast development of the horse tribe took place, so that great numbers of these animals left their remains in the lake-deposits of that time. The largest of these lakes had the Rocky Mountains for its western border, and extended from Dakota to Texas, its northern part covering the bed of the older Miocene basin. Another Pliocene basin, of unknown limits, extended over the older Tertiary strata of Eastern Oregon, and evidence of yet others may be seen in Idaho, Nevada, and California. In all of these fossil horses have been found; but the most important localities are the region of the Niobrara River east of the mountains, and the valley of the John Day River in Oregon (see sketch-map, p. 253).

It will thus be seen how abundant is the material for tracing the evolution of the horse in America. In that country the Equus fraternus of Leidy is believed to be almost, if not entirely identical with the Equus caballus (Linn.) of the old world.

Only in the upper Pliocene deposits does the true horse appear, and then the genealogy is complete. It roamed over the whole of North and South America, and soon after *seems* to have become extinct. There is no doubt that man and the horse were contemporaneous in early days; but it can be proved beyond doubt that, at the time of the Spanish conquest, few if any horses were left. It has been thought, from certain references contained in old narratives, that at least in South America the animal may possibly have still lingered on after the coming of the Europeans. What cause can have led to its extermination, it is impossible to say. The present race of wild horses, which roam in such vast herds on the Pampas, are not the descendants of the fossil horse of South America, but have sprung from those introduced by the Spaniards more than three hundred years ago.

A great deal of fresh material has been collected since the present writer, in the year 1894, published, in his *Creatures of Other Days*, some account of the evolution of the horse as told by the late Professors Marsh and Cope, and others; and some of the old conclusions have to be abandoned. Professor H. F. Osborn,[1] of the American Museum of Natural History, has taken up this line of research, and obtained important results. One of the trustees of the above museum, the late Mr. W. C. Whitney, generously provided funds for explorations. These expeditions began in the year 1890, and up to the year 1904, the American Museum Collection has been enriched by the more or less complete remains of 771 fossil horses, of which 146 were secured by Mr. J. W. Gidley, who took up this work under Professor Osborn. Five complete skeletons are now to be seen *mounted* in the above museum. All the other museums of the world together have only three complete mounted skeletons, and from all this wealth of material Professor Osborn has been able to find out three or four collateral lines of descent, some of which are mixed up with the pedigree published years ago by the late Professor Marsh.

According to Professor Osborn the line of evolution of the horse is, very briefly, as follows:—First we have the little Eohippus (or "dawn horse") only eleven inches (or two and three-tenths hands) at the withers, with its wrist or knee near the ground. The hand was still short, with four little hoofs,

[1] See his article in the *Century Magazine* for November, 1904, on the evolution of the horse in America (illustrated).

and part of a fifth. It had a wide range in America and Europe. The next stage is seen in the Hyracotherium (see Fig. 88), also

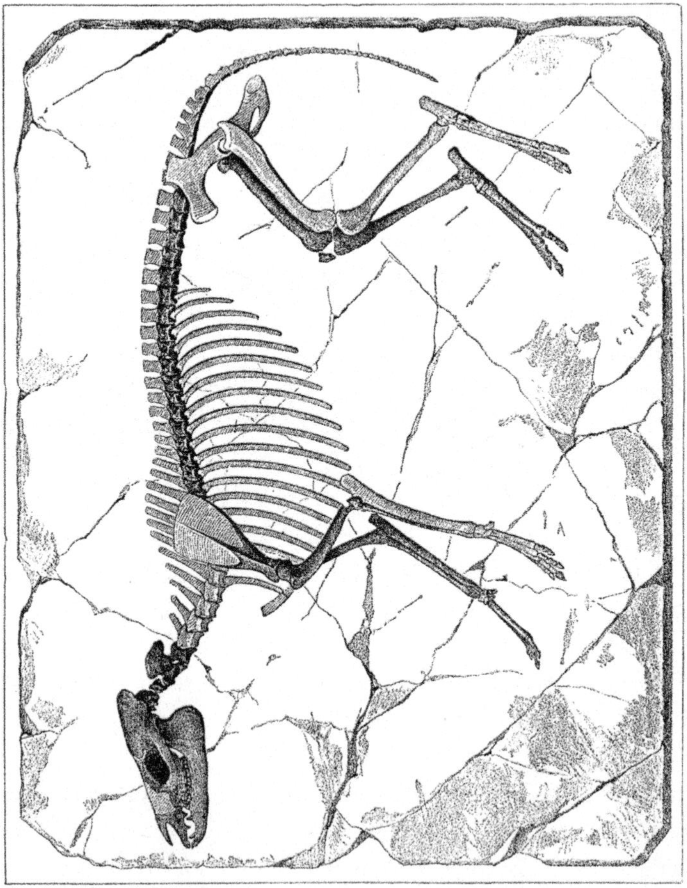

FIG. 88.—Skeleton of *Hyracotherium venticolum*, from Eocene strata, North America. (After Cope.)

known as Protorohippus. Its skull and forefoot are shown in Fig. 89. A nearly complete specimen was found by Dr. J. L. Wortman in the Big Horn Mountains of North Wyoming. It was a graceful little creature about fourteen inches (four and

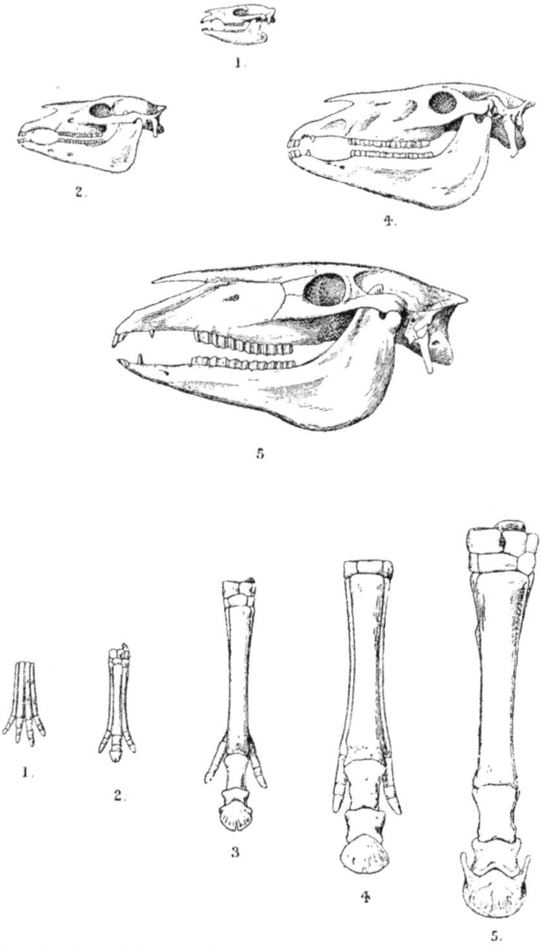

FIG. 89.—Skulls and bones of fossil horses showing evolution of the horse, from casts in the Natural History Museum, London.
1. *Protorohippus* (= *Hyracotherium.*) 2. *Mesohippus.* 3. *Protohippus.* 4. *Hipparion.* 5. *Equus caballus.*

a half hands) high at the withers. The next stage is seen in the Mesohippus (see Fig. 90). One finger has been dropped, and the skull is decidedly bigger. Here we see the middle or third digit getting larger. Unfortunately the skull of Protohippus, which is the next stage, is unknown; but we have the feet, of which a fore foot is shown in Fig. 89. Here the centre digit is much bigger, and this increase becomes still more striking in the Hipparion: until at last, in the modern horse, this one digit is supreme, and the others are only represented by the two feeble "splint bones" that were for so long a puzzle to anatomists. Now we see in them the vestiges of former digits no longer used for walking on. The outline restorations 1, 2, 3, 4, 5, 6, in Fig. 90, may be studied in connection with the bones in Fig. 89. Hipparion was of the size of a donkey.

South America had some peculiar horses of its own; for further information the reader should consult the *Guide to the Horse Family*, by Mr. R. Lydekker, F.R.S., to be obtained at the Natural History Museum (price one shilling).

Speaking of these changes, the late Sir William Flower said,[1] "Short, stout legs and broad feet, with numerous toes, spreading apart from each other when the weight of the creature is borne on them, are sufficiently well adapted for plodding deliberately over marshy and yielding surfaces, and the tapir and the rhinoceros, which in the structure of their limbs have altered but little from the primitive Eocene forms, still haunt the borders of streams and lakes and the shady depths of forests, as was probably the habit of their ancient representatives; while the horses are all inhabitants of the open plains, for life upon which their whole organisation is in the most eminent degree adapted. The length and mobility of the neck, position of the eye and ear, and great development of

[1] *The Horse: A Study in Natural History.* (London, 1891.)

1. *Hyracotherium* = *Protorohippus*.

2. *Orohippus*.

3. *Mesohippus* allied to the *Anchitherium*.

4. *Merychippus* (*Protohippus*).

5. *Pliohippus*.

6. *Equus caballus*.

FIG. 90.—The Evolution of the horse. 1–5, ancestors of the horse restored in outline from fossil remains from the Lower Eocene to recent times. (H. S. Lull, *American Jour. Sciencc*, vol. xxiii.)

the organ of smell, give them ample means of becoming aware of the approach of enemies; while the length of their limbs, the angles different segments form with each other, and especially the combination of firmness, stability, and lightness in the reduction of all the toes to a single one, upon which the whole weight of the body and all the muscular power are concentrated, give them speed and endurance surpassing that of almost any other animal."[1] At the same time that these changes were taking place (and in accordance with the "Law of Correlation") certain important changes in the structure and mode of growth of the teeth were being developed, the consequence of which was that they were able to masticate and grind up their food much more perfectly, and thus, no doubt, to obtain therefrom more vital energy with which to supply their muscles and nerves for working their limbs more rapidly as they became fitted to attain higher speed. They gained a grinding tooth for each toe that they lost. Of these ancient horses it may well be said they "ran the race that was set before them" or, perhaps we might say, which they set before themselves; and the victory was to the quickest runners—those who could "beat the record" made by their predecessors! It is a wonderful story, this, of the evolution of that noble and useful animal to which man is so much attached; and one cannot help sometimes wondering whether to some extent the *will* of an animal may not be an important factor in Evolution, although it is the fashion to ignore it, and to attribute organic changes to Natural Selection, or the "survival of the fittest." Mind has a powerful influence over matter, and can we not

[1] Visitors to the Natural History Museum will do well to examine the admirable set of casts and actual bones of fossil horses in Gallery No. 1. Complete casts of Phenacodus and Hyracotherium are there exhibited. In the hall may be seen the skeleton of a man and of a horse side by side, in the same relative position, also casts of bones represented in Fig. 89.

conceive that the earliest ancestors of the horse, finding (perhaps for the first time in the world's history) boundless grassy plains before them, were impelled by a strong desire to run?

It is a well-known fact that young horses are occasionally born with small extra toes on their feet. Suetonius[1] has given an account of the famous steed of Julius Cæsar. According to the historian, Cæsar "used to ride a remarkable horse, which had feet that were almost human, the hoofs being cleft like toes." This account clearly indicates the presence of extra digits, and the occurrence is more frequent than is generally supposed. Professor Marsh has made a study of this subject, since it has an important bearing upon the history of the evolution of the horse as given above; for in the light of modern teaching such occurrences are but "reversions" to older types that lived ages ago. He has therefore examined a large number of living animals with this peculiarity, and several interesting specimens of the same character have been sent to him: he has likewise received photographs, drawings, and detailed descriptions of various other examples, the authenticity of which cannot be questioned. The various cases he has come across may be summed up as follows:—

Sometimes there is only one extra digit on one foot. This is always much smaller than the main or third digit, the largest he has seen being about one-half its size, and the smallest very diminutive. This extra toe is almost invariably on the inner side of the main digit, and usually on the fore foot. Sometimes it is entirely hidden beneath the skin, the only external evidence of its presence being a prominence, which, on close examination, may be found to contain, below the "splint-bone," two or more bones corresponding to the finger-joints of our own hands. Sometimes a corresponding extra toe is developed on the other fore foot. At

[1] *De Vitâ Cæsaris*, lxvi.

other times a second extra digit has been found, as well as those above mentioned, which may be either smaller or larger.

Coming now to the hind feet, we find that an extra toe may sometimes be present on one or both of them, and always on the inside. As a rule, however, these posterior toes are much smaller, and often concealed beneath the skin. Fig. 91 is a drawing of a

Fig. 91.—Eight-toed Cuban horse. (After Marsh.)

horse from Cuba, in which the extra toe is well developed on all four limbs. This animal was examined by Professor Marsh during its lifetime.

Occasionally the hind feet have two extra toes, while the fore feet have only one, as in the horse from Texas, figured by Marsh;[1] but in rare cases both the fore and hind feet may each

[1] *American Journal of Science*, vol. xliii. p. 344.

have two extra toes fairly well developed, and all nearly of the same size. In such cases the result is something very like the feet of the ancient American type Protohippus, and the European Hipparion.

We have pointed out previously that the mammals of early Tertiary days had small brains, and the oldest ancestors of the horse are no exception to the rule. Professor Marsh has clearly proved that during this era a gradual increase in the size of the brain took place. It is interesting to find that the growth was mainly confined to the cerebral hemispheres, or higher portion of the brain. In most groups of mammals the brain has gradually become more convoluted, and thus increased in quality, as well as in quantity. In the long struggle for existence during the whole of Tertiary times the big brains won, as they do now. Applying this to our ancestral horses, it is easy to see that, as they acquired greater speed, and so roamed over larger tracts of country, they had to use their brains more.

We pass on now to give a brief account of a strange elephantine creature that lived in Eocene times, both in America and Europe, the Coryphodon[1] (so named from its teeth), a restoration of which is seen in Plate XL. The complete skeleton, as restored by Professor Marsh, is shown in Fig. 92.[2] The history of this remarkable animal, so long shrouded in obscurity, is worth recording here. The specimen on which the genus was founded by Sir R. Owen, in 1846, is unique, and was dredged up from the bottom of the sea, between St. Osyth and Harwich, and consists of the right branch of the lower jaw. This distinguished naturalist confessed that he had seldom felt more

[1] Greek—*korufé*, a ridge; *odous*, *odontos*, tooth.

[2] From Professor Marsh's paper, *Amer. Journ. Science*, xlvi. (1893), p. 325, a copy of which he kindly sent to the author.

misgiving in regard to a conclusion based on a single tooth or bone than that which he arrived at after a careful study of this specimen. Its smaller and less obvious features carried conviction to him against the showing of the larger and more catching ones. But although some naturalists for a time thought he had mistaken the fore for the back part of the jaw, yet his conclusion proved to be correct. His experience taught him that the less obvious points, which require searching out, frequently, when their full meaning has been grasped, guide to a right

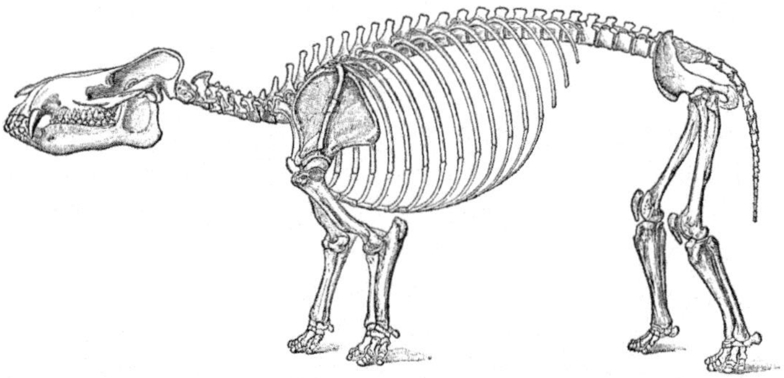

FIG. 92.—Skeleton of *Coryphodon hamatus*. (Restored, after Marsh.) Length about 6 feet.

interpretation of the whole. "It is as if truth were whispered," he says, "rather than outspoken by Nature."

The first additional evidence which Sir R. Owen obtained of the true nature of this ancient mammal was furnished by a fossil canine tooth brought up from a depth of a hundred and sixty feet out of the Plastic Clay during the operation of sinking a well in the neighbourhood of Camberwell, near London. This circumstance caused Sir R. Owen to remind his readers of the old proverb, "Truth lies at the bottom of a well." It was nearly three inches long, and evidently belonged to a large hoofed mammal. With regard to the teeth, he remarked that their broad-

CORYPHODON. EOCENE PERIOD OF EUROPE AND AMERICA.

Plate XL.

ridged and pointed grinding surfaces indicate that they were intended to be applied to the coarser kinds of vegetable substances.

According to Owen, certain fossils from the lignite deposits of Soisson, Laon, and Meudon, in France, belong to Coryphodon, Speaking of a tooth from Soisson, Cuvier said that the entire skeleton was found indicating an animal almost as large as a bull, but that the workmen employed in the sand-pit (*sablière*) unfortunately preserved only that one tooth. The first specimen of Coryphodon discovered in America was found in 1871, near Evanston, Wyoming, by Mr. William Cleburne, while engaged as surveyor for the Union Pacific Railroad, who secured chiefly its teeth and vertebræ. More or less perfect specimens were afterwards obtained by Cope during his explorations in New Mexico, under the survey of Captain Wheeler. Regarded merely as a fossil, it is characteristic of the Lower Eocene of Europe and America; in North America it is confined to the Wasatch and Wind River epochs, and is absent from the Upper Eocene of both countries.

Coryphodon is particularly interesting on account of the primitive features of its skeleton; the brain was very small and of a low type. Professor Cope places it, with its allied forms, in a separate order, to which he gives the name Amblypoda, on account of their elephantine limbs and probable ambling gait.[1]

C. hamatus was about six feet long; while some species were no larger than a tapir, others were as big as an ox (see Plate XL.).

Professor Cope thinks that in general appearance the Coryphodons resembled the bear more than any other living animal, with the important exception that in their feet they were like elephants.

[1] Mr. Charles Earle, after a careful revision of this family, has suggested that Cope's species should be greatly reduced in number, many of them in his opinion being only due to differences of age and sex. Bathmodon, Metalophodon, Ectodon, Manteodon may all be included in the genus Coryphodon.

The artist and author, however, in making the accompanying restoration, have been guided by the evident relationship between Coryphodon and the Tinoceras. "The movements of the Coryphodons," says Professor Cope, "doubtless resembled those of the elephant in its shuffling and ambling gait, and may have been even more awkward from the inflexibility of the ankle. But in compensation for the probable lack of speed, these animals were most formidably armed with tusks. These weapons, particularly those of the upper jaw, were more robust than those of the carnivora, and generally more elongate." There is no evidence that they had a proboscis—in fact, it is practically impossible. We may suppose from the nature of the teeth, and from other evidence, that they were vegetable feeders, but not restricted to any particular class of food; to a large extent they were omnivorous, like the hogs of to-day.

It is a little difficult to follow the curious interpretation arrived at by Professor Osborn and Dr. Wortman that "the positions of the fore and hind feet were absolutely different," the former being like those of the elephant (where only the tips of the toes touch the ground); the latter, in their opinion, like those of a bear and spreading out to rest on the ground (plantigrade).[1] This is not borne out by Professor Marsh's figure. Much valuable material for the study of the anatomy of this primitive mammal was collected by these two gentlemen, in spite of the many difficulties with which they had to contend, during their expedition, in the summer of 1891, into the Big Horn and Wind River regions, where the Wasatch strata are found.

The sketch-map shown in Fig. 93 will give the reader a general idea of the positions of the different geological

[1] *Bulletin of the American Museum of Natural History*, vol. iv. (1892), p. 121.

"basins"[1] in Western North America, where the strata have been found to yield so many valuable relics of ancient Tertiary life.

American geologists tell us that a long time ago (during the Eocene period) there was a great tropical lake in the Wyoming territory, on the borders of which roamed, amidst luxuriant

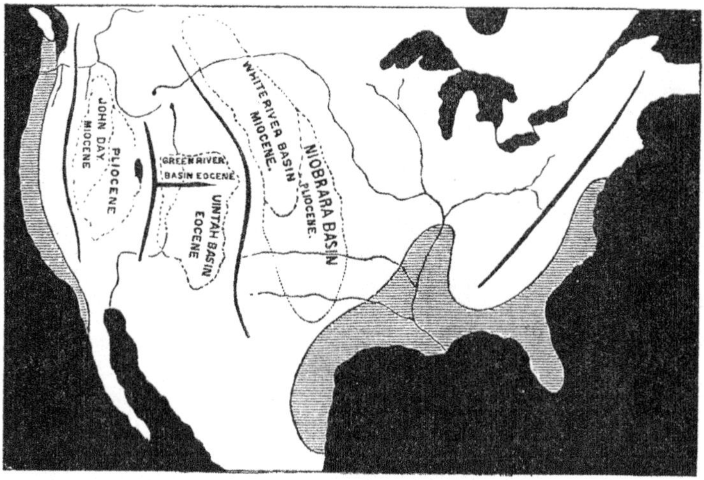

FIG. 93.—Map of North America in Tertiary times, showing roughly the outline of the coast, and sites of the principal fresh-water lakes.

vegetation, a large number of strange and primitive quadrupeds, together with many other forms of life. The most wonderful group of animals that haunted the shores of this lake, or perhaps river valley, was the Dinocerata so fully described by Professor Marsh, in his exhaustive monograph.[2] The name implies that

[1] When strata have been bent downwards into a kind of trough going far below the surface, they are said to lie in a "basin." Examples: the London basin, the Paris basin.

[2] *The Dinocerata*, a monograph by O. C. Marsh, *United States Geological Survey*, vol. x.

they were terrible horned monsters, but whether Nature provided them with true horns, like those of horned cattle to-day, is at least open to doubt.

Fig. 94 shows the skeleton of one of these, namely, Tinoceras ingens. Its length was about twelve feet without the tail. Its weight, when alive, was calculated to have been six thousand pounds, or about two tons and three quarters.

Plate XLI. is a restoration of the Tinoceras, made by our artist, after much consideration and careful study of the valuable cast

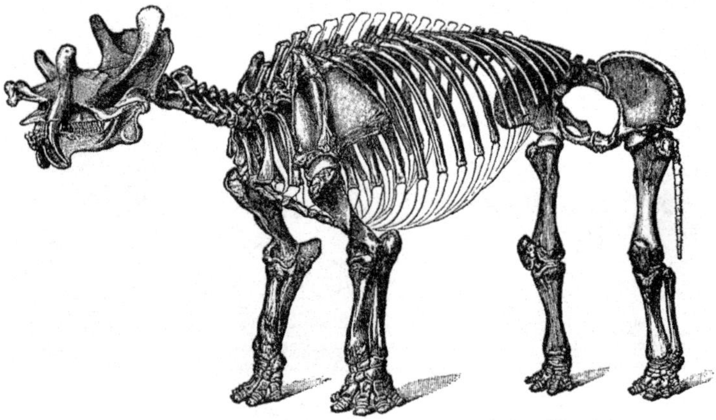

FIG. 94.—Skeleton of *Tinoceras ingens*. (After Marsh.)

exhibited in the Natural History Museum at South Kensington, which was generously presented by Professor Marsh.

Looking at the skeleton, one is struck with a certain resemblance to the rhinoceros on one hand, and to the elephant on the other. The legs are very elephantine, and the feet must have been covered with thick pads, but the body reminds one more of the rhinoceros. The eye was small and deep set, as in the rhinoceros. In the upper jaw the two canine teeth are developed into dagger-shaped tusks, the use of which can only be conjectured. In the females these are but slightly developed.

Plate XLI.

A LARGE EXTINCT MAMMAL, TINOCERAS INGENS. EOCENE PERIOD.
From North America. Length about 12 feet (without the tail).

It is quite clear, then, that we cannot place the Dinoceras in any order of living mammals. It is what palæontologists call a "generalised type;" that is to say, it presents certain characters seen in several groups of living quadrupeds, and not any of those elaborated or highly developed parts which we see in such animals to-day. Thus the proboscis of the elephant is the result of great elongation of the cheeks and nose of its ancestor (see

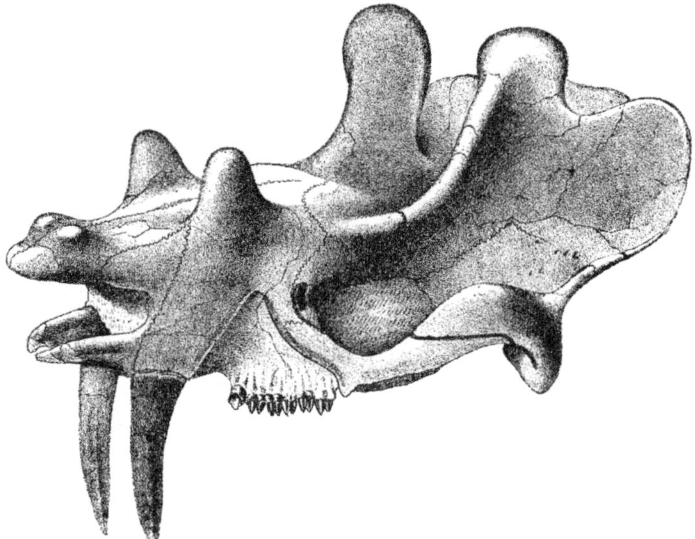

FIG. 95.—Skull of *Dinoceras mirabile*. (After Marsh.)

p. 269); in other words, the elephant is highly "specialised" in that direction, whereas our Dinoceras had no proboscis, or only a very slight one.

Again, the six remarkable bony protuberances of the skull served to some extent as horns, and probably were covered with thick bosses of skin, and did not support true horns like those of our modern oxen and other ruminants. Speaking of these protuberances, Professor Marsh says, "None of the covering of these elevations, or horn-cores, has, of course, been preserved; yet a fortunate discovery may perhaps reveal their nature by the form of a natural cast, as the eyeball of the Oreodon is sometimes thus clearly indicated in the fine Miocene matrix which envelops these animals." It looks rather as if we have here an early stage in the evolution of horns, and it may be that in the course of subsequent ages such prominences as those developed into true "horn cores," such as sheep or goats have, while the thick bosses of skin that covered them slowly developed into the true horns that are attached to these cores. If this is so, then we have here another instance of a "generalised" structure. Again, the limbs with their five toes tell us at once that the creature's place in Nature is outside of those two great groups of modern ungulates, or hoofed quadrupeds, the odd-toed and the even-toed, represented on the one hand by the horse, rhinoceros, and tapir, on the other by the pig, camel, deer, ox, and many other forms. Probably the two groups had not at this early period branched off from the primitive ungulate stock with five toes in each foot, of which the elephant is a living descendant, and from which also the Dinoceras must have come.

The limbs were strong and massive, but the brain was remarkably small, so that our Dinoceras cannot be credited with any high degree of intelligence: and here again we see an absence of "specialisation" compared with the sagacious elephant. Professor Marsh has taken casts of its brain-cavity (see Fig. 96). These casts show that the brain was smaller (in proportion to the size of the animal) than in any other mammal, whether living or extinct—and even less than in some reptiles! In fact, it was a decidedly reptilian kind of brain. Perhaps it may seem hardly credible, but so small was the brain of Dinoceras mirabile, that it could have been pulled through the apertures (neural canal) of all the neck vertebræ! In certain marsupials of the present day we find an approach to this kind of brain. It seems to be an established fact, according to Professor Marsh, that all the Eocene or earlier Tertiary mammals had small brains. His researches among fossil mammals have led him to the important conclusion that, as time went on, the brains of mammals grew larger; and thus he has been able to establish his law of brain-growth during the Tertiary period, a law which appears to be plainly recorded in the fossil skulls of succeeding races of ancient mammals. The importance of a discovery such as this cannot

FIG. 96.—Cast of brain-cavity of *Dinoceras mirabile*. (After Marsh.)

fail to strike the imagination of even the most unlearned in geology as being singularly suggestive and instructive. It is not difficult to picture these dull, heavy, slow-moving creatures haunting the forests and palm jungles around the margin of the great Eocene lake, into the waters of which their carcases from time to time found their way—perhaps swept down by floods. No footprints have been discovered as yet.

The Dinocerata were very abundant for a long time during the middle of the Eocene period. The position of their remains suggests that they lived together in herds, as cattle do now, and they probably found an abundance of food in the shape of succulent vegetation round the great lake. Geological evidence points to their sudden extinction before the close of the Eocene period; but it is difficult to understand this.

In the museum at Yale College are collected the spoils of numerous expeditions to the West, and the many tons of bones lying there are believed to represent the remains of no less than two hundred individuals of the Dinocerata. So perfectly have these bones been preserved by Nature that, even if the creatures had been living now, the material for studying their skeletons could hardly be more complete. Professor Marsh recognises three distinct types in this strange group of quadrupeds, on each of which a genus has been founded. The first and oldest form is the Uintatherium, which takes its name from the Uinta Mountains. This, as might be expected, is the most primitive or least specialised form, and comes from lower strata. The most highly developed or specialised form is the Tinoceras, and this is found at the highest geological level or "horizon."

Between these two extremes, and from an intermediate horizon, comes the Dinoceras,[1] so that in tracing these animals through the strata in which they occur the geologist finds that he is following for a while the course of their evolution. Doubtless there were many slight differences presented by the members of this group, but at present it has not been found possible to determine the number of species, although about thirty forms more or less distinct have been recognised. Professor Marsh says that the specimen of the skull of Dinoceras mirabile, on which the whole order Dinocerata was founded, is, fortunately, in a very perfect state of preservation, and that it belonged to a fully adult animal. Moreover, it was embedded in so soft a

[1] The Dincoeras of Marsh is the same form as Eobasileus of Cope. Uintatherium was discovered by Leidy.

matrix that the brain-cavity and the openings leading from it could be worked out without difficulty.

It is about forty-one years since the wonderful forms of life sealed up within these Eocene lake-deposits first became known to science. Long before then, however, the wandering Indian had been accustomed to seeing strange-looking skulls and skeletons that peeped out upon him from the sides of cañons and hills, as the rocks that enclosed them crumbled away under the influence of atmospheric agents of change—the ceaseless working of wind, rain, heat, and cold. To his untrained mind no other explanation suggested itself than the idea that these were the bones of his ancestors, which it would be highly impious to disturb. *Requiescant in pace!* So he left them in peace. Perhaps he believed in a former race of human giants; if so, these would be their bones. Long before Professor Marsh's expeditions, the earliest squatters, trappers, and others used to bring back news of marvellous monsters grinning from the ledges of rock beneath which they camped. At last these tales attracted the notice of some enthusiastic naturalists in the eastern States. Professor Leidy obtained a number of bones, from which he was able to bring to light an extinct creature at that time unknown to science, namely, the Uintatherium. Professor Cope also described some extinct animals disinterred by himself from the same region.

But our knowledge of the Dinocerata is chiefly due to Professor Marsh, who has despatched one expedition after another, and who, after many years of laborious research both in the western deserts and in his wonderful collection at Yale College, has published a splendid monograph on the subject. No trouble and no expense have been spared in order to obtain material for this great work, and all geologists must feel grateful to Professor

Marsh for so liberally devoting his time and his private resources in order to advance the science of Palæontology.

In the country east of the Rocky Mountains, including the states of Dakota, Nebraska, Wyoming, and part of Colorado, Professor Marsh has discovered the remains of yet another strange group of large quadrupeds. The best known of these is Brontops, of which the skeleton is seen in Fig. 97. These animals lived after the Dinocerata, namely, in the Miocene period, and

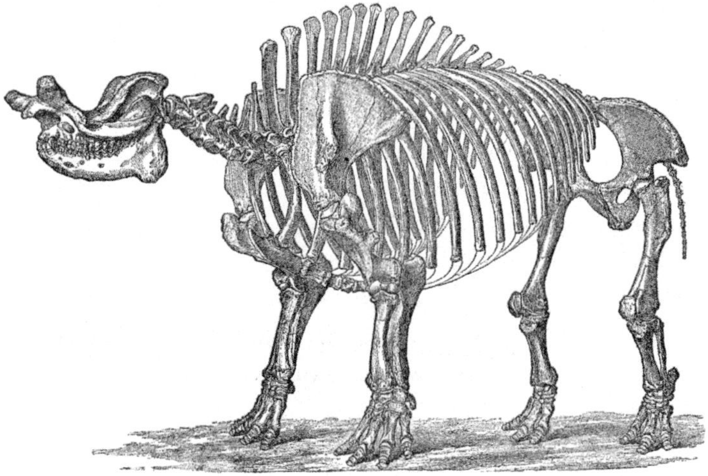

FIG. 97.—Skeleton of *Brontops robustus*. (After Marsh.)

were the largest American mammals of that period. They constitute a distinct family more nearly allied to the rhinoceros than to any other living form. The skeleton on which Fig. 97 is founded was the most complete of any yet discovered by Professor Marsh. Portions of it were exhumed at different times, but it was first found in 1874. Our artist has made the restoration seen in Plate XLII. from this skeleton, as figured by Professor Marsh.

This strange group of creatures flourished in great numbers on

PLATE XLII.

A HUGE EXTINCT MAMMAL, BRONTOPS ROBUSTUS, FROM NORTH AMERICA. MIOCENE PERIOD.

the borders of an old lake of Miocene age. The Brontops was a heavy, massive animal, larger than any of the Dinocerata, with a length of twelve feet, not including the tail, and a height of eight feet. The limbs are shorter than those of the elephant, which it nearly equalled in size. As in the tapir, there were four toes to the front limbs, and three to the hind limbs. Its skull was of

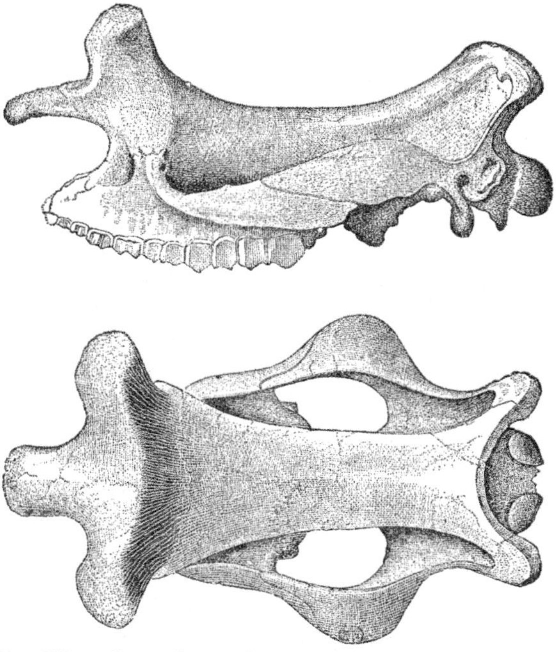

FIG. 98.—Side and top views of skull of *Titanotherium*, from Miocene strata, North America.

a peculiar shape, shallow, and very large. That of Brontops ingens is thirty-six inches long, and twenty inches between the tips of the two horns, or protuberances. The creature was probably provided with an elongated, flexible nose, like that of the tapir, but not longer, because the length of the neck shows that it could reach the ground without the aid of a trunk such as the elephant's. This creature has also been described under the name of Titanotherium. (See Plate XLIII.)

One of the strangest of quadrupeds ever discovered by geologists was the huge and fierce beast known as Arsinoëtherium of which

FIG. 99.—Skull and lower jaw of *Arsinoëtherium*, from the Upper Eocene of the Fayûm, Egypt.

the skull is shown in Fig. 99, and a restoration in Plate XLIII. The remains were found by Mr. Beadnell of the Egyptian Geological Survey in the Upper Eocene Sands of the Fayûm [1] district. The only parts missing appear to be the ribs and the toes. The skull was nearly three feet long, and although some

[1] See *Catalogue of the Tertiary Vertebrata of the Fayûm*, Egypt. By C. W. Andrews, D.Sc. Published by Trustees of British Museum.

PLATE XLIII.

NOT A RHINOCEROS, ARSINOËTHERIUM ZITTELI. EOCENE AGE. EGYPT.
Length about 9 feet.

bones are still missing, enough is known to justify a tentative restoration such as is shown in our Plate XLIII. See the specimens in the Fossil Mammal Gallery, Natural History Museum. Dr. C. W. Andrews proposes a new order, viz. the Barypoda (or heavy-footed beasts), to distinguish this creature from such of its allies as the Dinoceras and the Elephants. The affinities of Arsinoëtherium are at present unsettled, but it may possibly be related to the order of ungulates known as Hyracoidea, and at the present day represented by the Syrian Hyrax or coney. It is an isolated and a primitive group, probably related to the Phenacodus, as well as to the order Amblypoda. Queen Arsinoë, a Greek Queen of Egypt, had her palace near where the bones were found; hence the name. The horn-cores in the skull are formed by a hollow shell of bone. The height of the creature was 5 feet 9 inches at the withers, and the length about 9 feet 9 inches from the snout to the rump.

CHAPTER XIV

THE STORY OF THE ELEPHANTS

"Yes, where the huntsman winds his matin horn,
 And the couched hare beneath the covert trembles;
Where shepherds tend their flocks, and grow their corn,
 Where fashion in our gay Parade assembles—
Wild horses, deer, and elephants have strayed,
 Treading beneath their feet old Ocean's races."

HORACE SMITH.

THE science of Palæontology is always advancing, sometimes by leaps and bounds, at other times slowly but surely. It is our pleasant duty to record here a great step in advance made some seven years ago chiefly by the researches in Egypt of Dr. C. W. Andrews, F.R.S. His discoveries in the district known as the Fayûm have enabled him to solve a very important and interesting problem, namely, the evolution of the Elephant, the only living representative of that strange order known to naturalists and palæontologists as the Proboscidia, because they are all provided with trunks. The relationships of the group are still partly wrapped in mystery,[1] but as if to compensate for this loss, we can now tell how the elephant got his trunk! We can almost see it growing as we look at those wonderful skulls obtained from Egypt, showing, among other things, the great increase in size which took place. This is surely a great result, the problem of

[1] According to Dr. C. W. Andrews, and others, they seem to be related to the Sirenia or sea-cows, an order very unlike in external appearance, but for all that the anatomist finds certain common characters.

THE EVOLUTION OF THE ELEPHANT. AN EARLY ANCESTOR, MŒRITHERIUM. EOCENE PERIOD. EGYPT.

PLATE XLIV.

Height about 3 feet at the shoulder.

Illustration lent by H. R. Knipe (from *Nebula to Man*).

the evolution of the Proboscidia is solved at last! It will not be necessary to go into many details in our attempt to explain this wonderful story. If the reader should go to the Natural History

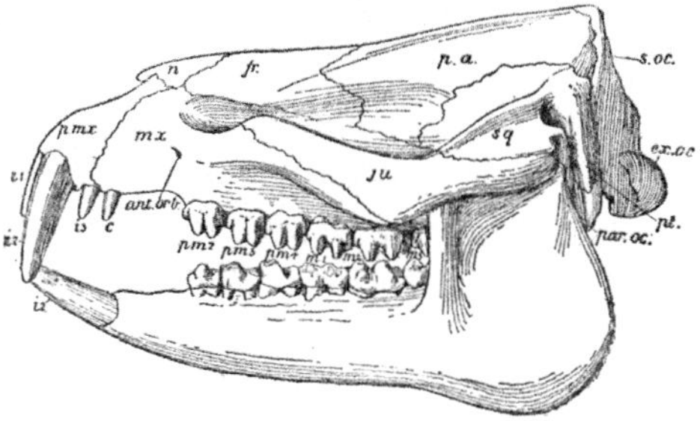

FIG. 100.—Skull and lower jaw of *Mœritherium lyonsi*, Middle Eocene of the Fayûm, Egypt. (After C. W. Andrews.)

Museum at South Kensington he will be able to inspect the models of restored skulls which have been made at the Museum.

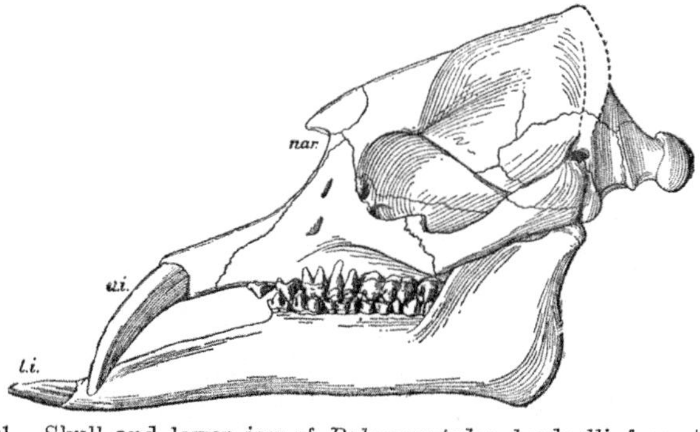

FIG. 101.—Skull and lower jaw of *Palæomastodon beadnelli*, from the Upper Eocene of the Fayûm, Egypt. (After C. W. Andrews.)

These are to be seen close to the elephant skeletons in the gallery devoted to fossil mammals.

The four stages in the history of Elephants may be best illustrated by the four skulls shown in Figs. 100 to 103. The first belongs to the Mœritherium, which takes its name from the old

Egyptian Lake Mœris. "Can it be possible," the reader may well ask, "that this skull once belonged to a creature which was the forerunner of Elephants?" Well, so it seems, although the differences are very great. The shape of the skull gives no suggestion at all of anything elephantine, nor does the large number of teeth with their sharp pointed ridges. The Mœritherium, according to Dr. Andrews,[1] was about the size of a tapir, which it much resembled, and it probably frequented the marshes of the period

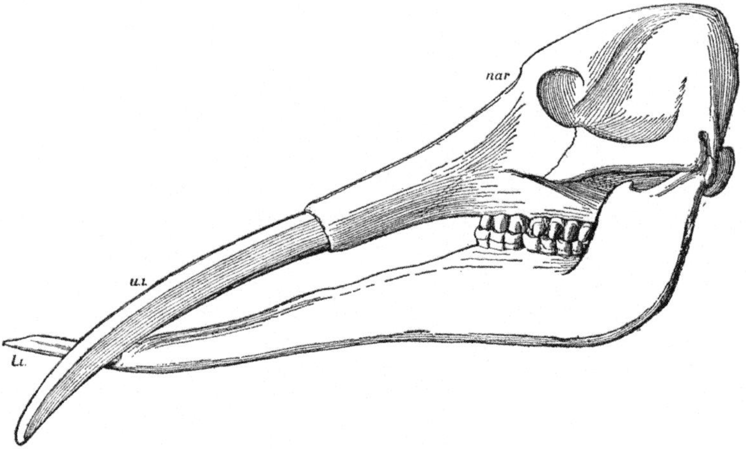

FIG. 102.—Skull and lower jaw of *Tetrabelodon augustidens*, from Middle Miocene of Sansan, France. (After C. W. Andrews.)

(Middle Eocene), see Plate XLIV. The length of the skull is only about sixteen inches; the limbs are partly known. The next beast, the Palæomastodon beadnelli was larger and more adapted to a terrestrial life. Its skull, shown in Fig. 101, was about three feet long, and now we see something really elephantine. The whole face, nose, and cheeks began to be pushed forward, its forehead slopes upward, and there is a large opening for the nose, the teeth

[1] *Catalogue of the Tertiary Vertebrata of the Fayûm Egypt.* Published by Trustees of the British Museum.

THE EVOLUTION OF THE ELEPHANT, SECOND STAGE, PALÆOMASTODON. EOCENE PERIOD. EGYPT.

PLATE XLV.

Height about 4 to 6 feet at the shoulder.

Illustration lent by H. R. Knipe (from *Nebula to Man*).

are reduced in number. The canines and the first premolars are lost. The lower jaw is considerably elongated. We begin to see signs of a trunk. Now, this creature probably looked like a rather long-necked elephant (see Plate XLV.), but its trunk was not flexible, being supported by the long lower jaw, for which it partly formed a covering. Probably it could reach the ground with its lower incisors, and the end of the snout may have been prehensile. In size it was about as big as a half-grown Indian Elephant. Now, in the next stage, represented by Tetrabelodon augustidens

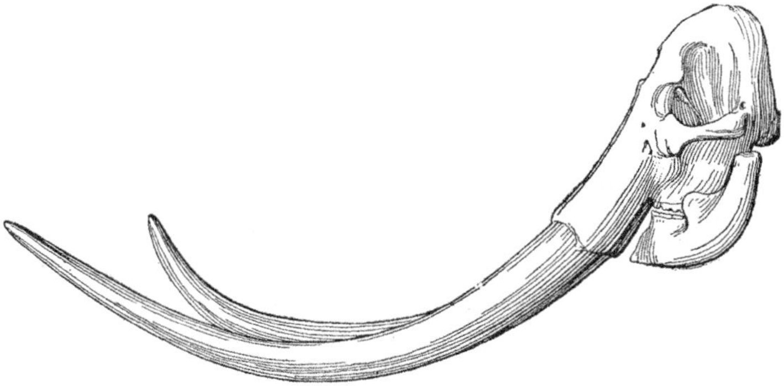

FIG. 103.—Skull and lower jaw of *Elephas* (*Stegodon*) *ganesa*, from Lower Pliocene of Sivalik Hills, N. India.

(see Fig. 102), we have something much more like an Elephant. To begin with, it is bigger. Then see how much more elongated is the lower jaw. The upper incisors have greatly increased in size; only two molars are left in each jaw. The opening for the nose is further back. Plate XLVI. is a restoration. In size this creature also resembled the Elephant, being about as big as a moderate-sized elephant of the present day. In the later Miocene period this long lower jaw got shortened, and so an approach was made to the modern type. Fig. 103 shows us the last stage in this strange history, when the lower jaw rather suddenly contracted and the proboscis greatly increased in size, being left entirely free and unsupported. These remarkable changes have been well summed up by Dr. A. Smith Woodward, F.R.S., in the following words:—"They gradually increased in size without essentially altering their limbs and body; but as their legs lengthened and their neck shortened, their face and chin gradually became elongated to reach the ground for browsing. When this strange adaptation had reached its maximum degree, the chin suddenly shrivelled, leaving the flexible, toothless face without any support.

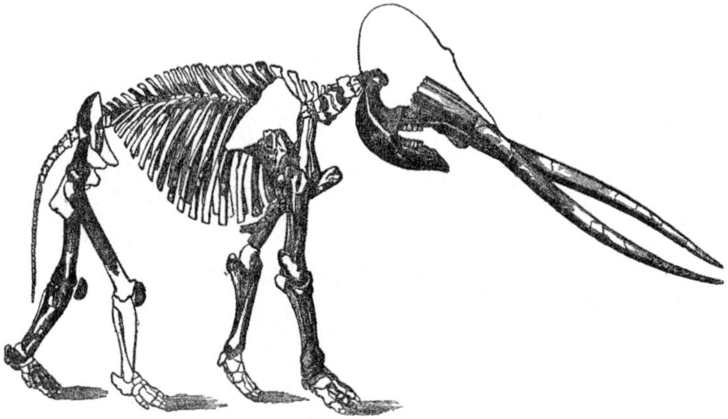

FIG. 104.—Skeleton of *Mastodon arvernensis*, Pliocene, Europe.

Thus arose the unique proboscis of the Elephants, which has become prehensile by stages which cannot be traced, because soft parts are not preserved in ordinary geological formations."

During the Miocene period there lived a strange kind of elephant known as the Dinotherium. No complete skeleton is known, but the skull, limb-bones, vertebræ, etc., have been found. The skull was discovered at Eppelsheim, in Hesse-Darmstadt, and other remains in France, Greece, and Asia Minor, also closely allied forms in India, but none in America. We learn from the

THE EVOLUTION OF THE ELEPHANT, THIRD STAGE, TETRABELODON.

MIOCENE PERIOD. EGYPT.

PLATE XLVI. Height about 7 or 8 feet at the shoulder.

Illustration lent by H. R. Knipe (from *Nebula to Man*).

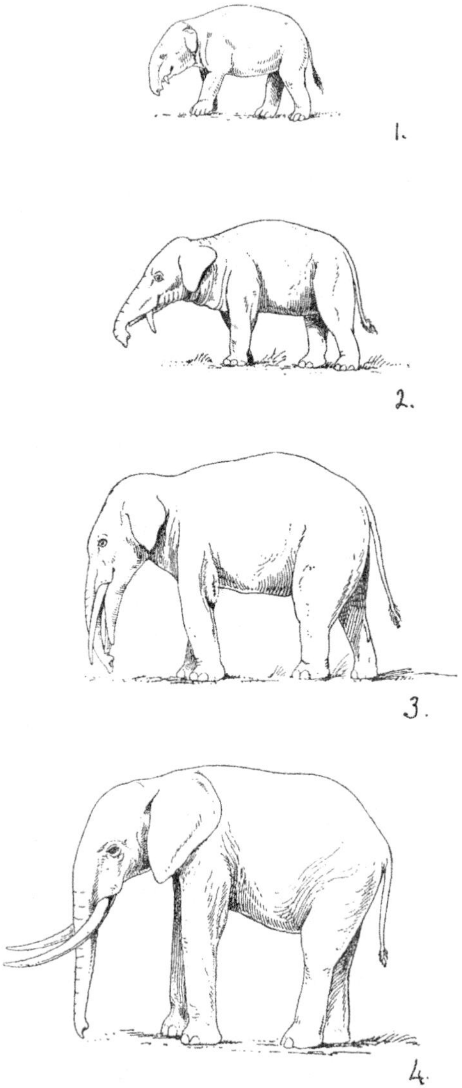

FIG. 105.—Evolution of the Elephant.

1. *Mœritherium.* 2. *Palæomastodon.*
3. *Tetrabelodon.* 4. *Elephas africanus.*

skull of this creature that it had a fairly large proboscis, but, unlike other elephants of the present time, the two tusks are in the lower jaw and curve downwards instead of upwards. Dean Buckland came to the conclusion that this creature was largely aquatic in its habits, but this is not certain. However, the aquatic habits of the tapir, a distant relation of the elephant, show that this conclusion is by no means an impossible one. It is not easy to imagine how such tusks were used under water, as Buckland suggested, to grub up aquatic plants; but on land they could certainly be used for raking up the ground, and perhaps in pulling down the branches of trees. A length of eighteen feet has been attributed to this creature, of which the skull is shown in Fig. 106, but this is probably an over-estimate. A restoration is shown in Plate XLVII.

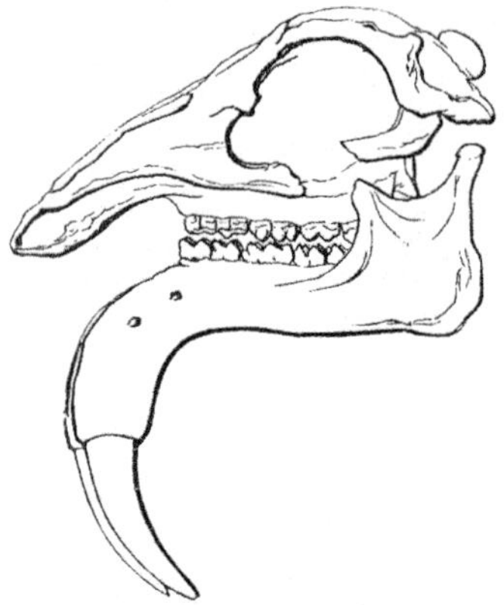

FIG. 106.—Skull of *Dinotherium giganteum*, from Eppelsheim. The lower jaw is a cast. An examination of the skull at the top will show that it is considerably crushed in.

THE MAMMOTH

Many are the traditions and tales that have clustered round the Mammoth.[1] He is, however, no fabulous product of the

[1] The word *Mammoth* is thought by Pallas and Nordenskiöld to be of Tartar origin. The former asserts that the name originated in the word *mamma*, which signifies earth (the Mammoth being found frozen in the earth). It was introduced into the languages of Western Europe about two centuries ago, from the Russian. But other writers have attempted to prove that it is a corruption of the Arabic word *Behemoth*, or "great beast," which in the Book

AN ANCIENT ELEPHANTINE MONSTER, DINOTHERIUM. MIOCENE PERIOD. EUROPE.

PLATE XLVII.

imagination, like the dragon, for he has actually been seen in the flesh, and not only *seen*, but eaten, both by men and animals! But, for all that, men's minds have been busy for centuries past making up tales, often of the wildest description, about him; and it is little wonder that a creature whose bones are found in the soils and gravels, etc., over more than half the world, and whose body has been seen frozen in Siberian ice, should have given rise to many tales and superstitions. To students of folk-lore these legends are of considerable interest, and to some extent also to men of science. To the latter, however, one of its many points of interest is that palæontology may be said to have been founded on the Mammoth. Cuvier, the illustrious founder of the science of organic remains, was enabled, by his accurate and minute knowledge of the structures of living animals, to prove to his astonished contemporaries that the Mammoth bones and teeth, so plentifully discovered in Europe, were not such as could have belonged to any living elephant, and consequently that there must have existed, at some previous period in the world's history, an elephant of a different kind, and quite unknown to naturalists. This was a new idea, and accordingly one that met with opposition as well as incredulity.

It was thought in those days that whatever animals lived in the past *must* have resembled those now inhabiting the world, and the idea of extinct types unknown to man, and unknown to the regions where their bones were found embedded below the soil, was of so novel and startling a character as to appear incredible. Besides, the Mosaic account of Creation made no direct reference to extinct animals, and therefore the notion was not to be entertained!

of Job signifies an unknown animal. In an ancient Chinese work, of the fifth century before Christ, it is spoken of under the name *tien-schu*, that is to say, "the mouse which hides itself."

It is amusing to note the devices to which people resorted in order to combat this revolutionary teaching. Thus, when Cuvier first announced the discovery of the fossil remains of the elephant, hippopotamus, and rhinoceros in the superficial deposits of continental Europe, he was gravely reminded of the elephants introduced into Italy by Pyrrhus in the Roman wars, and afterwards in the Roman triumphal procession or the games at the Colosseum.

It was only by means of minute anatomical differences that he was able to show that the bones and teeth of these elephants must have belonged to a species unlike those now living. But these differences proved too subtle for even scientific men to appreciate, so slight was their knowledge of anatomy compared with his; so that they were either disallowed or explained away.

But he was not to be beaten, and appealed to the fact that similar remains occurred in Great Britain, whither neither Romans nor others could have introduced such animals. These are his words: "If, passing across the German Ocean, we transport ourselves into Britain, which in ancient history by its position could not have received many living elephants besides that one which Cæsar brought thither, according to Polycenus; we shall, nevertheless, find these fossils in as great abundance as on the Continent."

Another crushing answer to the absurd explanations of Cuvier's countrymen was added by the sagacious Dean Buckland, who pointed out that in England, as on the Continent, the remains of elephants are accompanied by the bones of the rhinoceros and hippopotamus, animals which not even Roman armies could have subdued or tamed! Owen also adds that the bones of fossil elephants are found in Ireland, where Cæsar's army never set foot.

It was in 1796 that Cuvier announced that the teeth and bones of the European fossil elephants were distinct in species from both the African and the Indian elephant, the only two living species (El. africanus and El. indicus). This fundamental fact opened out to him new views about the creation of the world and its inhabitants, and a rapid glance over other fossil bones in his collection showed him the truth and the value of this great idea (namely, the existence of extinct types), to which he consecrated the rest of his life. Thus palæontology may be said to have been founded on the Mammoth.

The fossil remains of elephants have, on account of their common occurrence in various parts of the world, attracted a great deal of attention, both from the learned and the unlearned. In the north of Europe they have been found in Ireland, in Germany; in Central Europe, in Poland, Middle and South Russia, Greece, Spain, Italy; also in Africa, and over a large part of Asia. In the New World they have been found abundantly in North America. But in the frozen regions of Siberia their tusks, teeth, and bones are met with in very great abundance. According to Pallas, the great Russian savant, there is not in the whole of Asiatic Russia, from the Don to the extremity of the Tchutchian promontory, any brook or river on the banks of which some bones of elephants and other animals foreign to these regions have not been found. The primæval elephants (Mammoth, Mastodon, etc.) appear to have formerly ranged over the whole northern hemisphere of the globe, from the fortieth parallel to the sixtieth, and possibly to near the seventieth degree of latitude.

Just as the North American Indian regards the great bones of Professor Marsh's extinct Eocene mammals that peep out from the sides of buttes and cañons, as belonging to his ancestors, so we find that in all parts of the world the bones of extinct elephants have, on account of their great size (and partly from a certain resemblance, in some, to bones of the human skeleton), been regarded as testifying to the former existence of giants, heroes, and demigods. To the present day the Hindoos consider such remains as belonging to the *Rakshas*, or Titans,—beings that figure largely in their ancient writings. Theophrastus, of Lesbos, a pupil of Aristotle, appears to have been the first to record the discovery of fossil ivory and bones. These were probably obtained by the country people from certain deposits in the neighbourhood, and are mentioned five hundred years later by Pausanias. Several Greek legends and traditions appear to be founded on such discoveries.

We will now put before the reader the true accounts given by Adams.[1]

In 1799 a Tungusian, named Schumachoff, who generally went to hunt and fish at the peninsula of Tamut after the fishing season of the Lena was over, had constructed for his wife some cabins on the banks of the lake Oncoul, and had embarked to seek along the coasts for Mammoth tusks. One day he saw among the blocks of ice a shapeless mass, but did not then discover what it was. In 1800 he perceived that this object was more disengaged from the ice, and that it had two projecting parts; and towards the end of the summer of 1801 the entire side of the animal and one of his tusks were quite free from ice. In 1803 the enormous mass fell by its own weight on a bank of sand. It was a frozen Mammoth! In 1804 Schumachoff came to his Mammoth, and having cut off the tusks, exchanged them with a merchant for goods. Two years afterwards Mr. Adams, the narrator of the story, traversed these

[1] Abridged from *Memoirs of the Imperial Academy of Sciences of St. Petersburg*, vol. v. London, 1819.

REMAINS OF A MAMMOTH EXCAVATED BY THE MEMBERS OF AN EXPEDITION FROM ST. PETERSBURG IN 1902 ON THE BANKS OF THE BEROSOWKA (JAKUTSK).

From a paper by Dr. Hertz, Bulletin Imp. Academy, St. Petersburg, April, 1902, old style.

PLATE XLVIII.

distant and desert regions, and found the Mammoth still in the same place, but sadly mutilated. The people of the neighbourhood had cut off the flesh, and fed their dogs with it during the scarcity. Wild beasts, such as white bears, wolves, and foxes, also had fed on it, and the traces of their footsteps were seen around. The skeleton was complete all except one leg, but the flesh had almost all gone. The head was covered with a dry skin, one of the ears was seen to be covered with a tuft of hairs. All these parts suffered more or less injury in transport for a distance of 7330 miles to St. Petersburg, yet the eyes have been preserved. This Mammoth was a male, with a long mane on its neck, but both tail and proboscis had disappeared. The skin is of a dark grey colour, covered with a reddish wool and black hairs. The entire carcase was nine feet four inches high. The skin of the side on which the carcase had lain was detached by Mr. Adams, for it was well preserved, but so heavy was it that ten persons found great difficulty in transporting it to the shore. The white bears, while devouring the flesh, had trodden into the ground much of the hair belonging to the carcase, but Mr. Adams was able by digging to procure about sixty pounds' weight of hair. In a few days the work was completed, and he found himself in possession of a treasure which amply compensated him for the fatigues and dangers of the journey as well as the expense of the enterprise. When first seen, this Mammoth was embedded in clear pure ice, which forms in that coast escarpments of considerable thickness, sloping towards the sea, the top of which is covered with moss and earth. If the account of the Tungusians can be trusted, the carcase was some way below the surface of the ice when first seen. Arrived at Takutsk, Mr. Adams purchased a pair of tusks which he believed to belong to this Mammoth, but there is reason to doubt whether he did get the right tusks. They are nine feet six inches long.

The skeleton of this specimen, the fame of which may be said to have spread all over the world, is now set up in the Museum of the St. Petersburg Academy, and the skin still remains attached to the head and feet. A part of the skin and some of the hair were sent by Mr. Adams to Sir Joseph Banks, who presented them to the Museum of the Royal College of Surgeons.[1] A photograph of the skeleton as it now stands, may be seen on the wall of the Geological Gallery at South Kensington near the specimens of Mammoth tusks. But it should be pointed out that *the tusks are put on the wrong way;* for they curve outwards instead of inwards, thus presenting a somewhat grotesque appearance. For this reason we have not reproduced the familiar woodcut based on an engraving in the memoir already referred to.[2] But we give, instead, a sketch taken from a photograph (also on the wall in Gallery No. 1.) of a fine skeleton in the Brussels Museum (Fig. 107). Here the tusks are seen correctly inserted. We must also draw the reader's attention to the remarkably fine specimen consisting of the skull and both tusks complete, found at Ilford in Essex. Plate XLVIII. shows a more recent discovery in Siberia.

Adams's specimen was, Dr. H. Woodward thinks, an old individual, and its tusks had curved upwards so much as to be

[1] A specimen of the hair of a mammoth may be also seen at the Natural History Museum in a tall glass jar. It came from frozen soil, Behring Strait. By the side of this will be seen, in a glass box, a portion of the skin of a mammoth, from the banks of the river Alaseja, Province of Yakutsk, Siberia. It exhibits the under fur, the long hair having entirely disappeared.

[2] Fig. 42 in the *Guide to the Fossil Mammals and Birds in the Department of Geology and Palæontology in the British Museum (Natural History), Cromwell Road*, 1904. This most useful guide should be consulted by the reader. Also *A Guide to the Elephants*, 1908. Each 6*d*.

of little use. In younger ones they were less curved. The hair that still remains on the skin of the St. Petersburg specimen is of the colour of the camel, very thick-set and curled in locks. Bristles of a dark colour are interspersed, some reddish, and some nearly black. The colour of the skin is a dull black, as in living elephants (see restoration, Plate XLIX.).

Remains of the Mammoth (Elephas primigenius) have been found in great numbers in the British Isles. Mr. Samuel

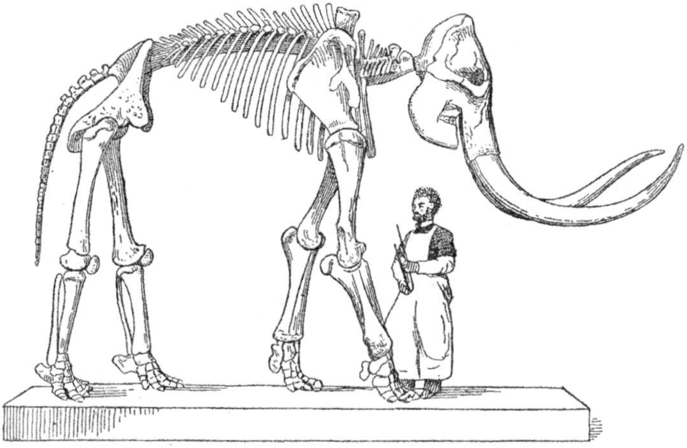

FIG. 107.—Skeleton of Mammoth, *Elephas primigenius* (partly restored), in the Museum at Brussels. Drawn (from a photograph) by J. Smit.

Woodward calculated that upward of two thousand grinders of elephants have been dredged up during a period of thirteen years upon the oyster-bed off Hasborough, on the Norfolk coast. But many of these doubtless belong to other species of older date, such as Elephas antiquus.

Dr. Bree, of Colchester, says that the sea-bottom off Dunkirk, whence he has made a collection, is so full of mammalian remains that the sailors speak of it as "the Burying-ground."

The remains of the Mammoth occur over a very large geographical area—fully half the globe.

By far the most important discovery of a frozen Mammoth is that of a young Russian Engineer, Benkendorf by name, who was an eye-witness of its resurrection, though, most unfortunately, he was unable either to procure his specimen, as Mr. Adams did, or to make drawings of it. A full account was given in our previous edition, p. 204.

With regard to the food of the Mammoth, Benkendorf's discovery is of great service in solving the question how such a creature could have maintained its existence in so inhospitable and unpromising a country. The presence of fir-spikes in the stomach is sufficient to prove that it fed on vegetation such as is now found at the northern part of the woods as they join the low treeless tundra in which the body lay buried.

Before this discovery the food of the Mammoth was unknown, and all sorts of theories were devised in order to account for its remains being found so far north. Some thought that the Mammoth lived in temperate regions, and that the carcases were swept down by great floods into higher and colder latitudes. But it would be impossible for the bodies to be hurried along a devious course for so many miles without a good deal of injury, and probably they would fall to pieces on the way. But, as Professor Owen has so convincingly argued, there is no reason why herds of Mammoths should not have obtained a sufficient supply of food in a country like the southern part of Siberia, where trees abound in spite of the fact that during a great part of the year it is covered with snow. And this is his line of reasoning. The molar teeth of the elephant show a highly complicated and peculiar structure, and there are no other quadrupeds that feed to such an extent on the woody fibre of the

THE MAMMOTH, ELEPHAS PRIMIGENIUS.

From a drawing by Charles R. Knight. From *Animals of the Past*, by Prof. F. A. Lucas, by permission of Messrs. Doubleday, Page and Co., New York.

PLATE XLIX.

branches of trees. Many mammals, as we know, eat the leaves of trees; some gnaw the bark; but elephants alone tear down and crunch the branches. One would think there was but little nourishment to be got from such. But the hard vertical plates of their huge grinders enable them to pound up the tough vegetable tissue and render it more or less palatable. Of course, the foliage is the most tempting, but where foliage is scarce something more is required.

Now, in the teeth of the Mammoth the same principle of construction is observed, only with greater complexity, for there are more of these grinding plates and a larger proportion of dense enamel. Hence the inference seems unmistakable that the extinct species fed more largely on woody fibre than does the elephant of to-day. Forests of hardy trees and shrubs still grow upon the frozen soil of Siberia, and skirt the banks of the Lena as far north as the sixtieth parallel of latitude.

If the Mammoth flourished in temperate latitudes only, as formerly suggested, then its thick shaggy coat becomes superfluous and meaningless; but if it lived in the region where its body has been found, then the argument from its teeth, and the fir-spikes found in its stomach, is confirmed by the nature of its skin, and all the old difficulties vanish. Sir Richard Owen considered that we may safely infer that, if living at the present day, it would find a sufficient supply of food at all seasons of the year in the sixtieth parallel, and even higher. Perhaps it migrated north during the summer; and, judging from the present limits of arboreal vegetation, it may have been able to subsist even in latitude 70° north, for at the extreme points of Lapland pines attain a height of sixty feet.[1]

[1] Sir Henry Howorth, in his *Mammoth and the Flood*, suggests another theory, and gives much valuable information.

It is often no easy matter to form conclusions with regard to the habits of extinct animals; and we must not rely too much on arguments derived from the habits of their living descendants or relatives. The older geologists fell into this mistake, as did even Cuvier. Modern elephants are at present restricted to regions where trees with perennial foliage flourish; and therefore it was argued that there must have been a change of climate, either gradual or sudden in the country of the mammoth. The late Sir Richard Owen, however, did not share these views, as the following extract shows: "The fact seems to have been generally overlooked that an animal organised to gain its subsistence from the branches or woody fibre of trees, is thereby rendered independent of the seasons which regulate the development of leaves and fruit; the forest food of such species becomes as perennial as the lichens that flourish beneath the winter snows of Lapland; and were such a quadruped to be clothed, like the reindeer, with a natural garment capable of resisting the rigours of an arctic winter, its adaptation for such a climate would be complete. . . . The wonderful and unlooked-for discovery of an entire Mammoth, demonstrating the arctic character of its natural clothing, has, however, confirmed the deductions which might have been legitimately founded upon the localities of its most abundant remains, as well as upon the structure of its teeth, viz. that, like the Reindeer and Musk Ox of the present day, it was capable of existing in high northern latitudes." [1]

The problem of the extinction of the Mammoth is not an easy one to solve. We can hardly account for its disappearance by calling in geographical changes by which its range became restricted and its food supply diminished, so that in the

[1] *A History of British Fossil Mammals and Birds*, by Richard Owen, F.R.S., etc. London, 1846.

competition with other herbivorous animals this primæval giant "went to the wall," as the saying is. Nor does Lyell's appeal to a change in climate, by which the cold of Siberia became too intense even for the Mammoth, seem quite satisfactory, especially when we remember how very far north fir trees range (p. 279).

The Mammoth, probably, was endowed with a fairly tough constitution. See Plate XLIX. In Siberia it fed on fir trees. In Kentucky it fared better, and was surrounded by such vegetation

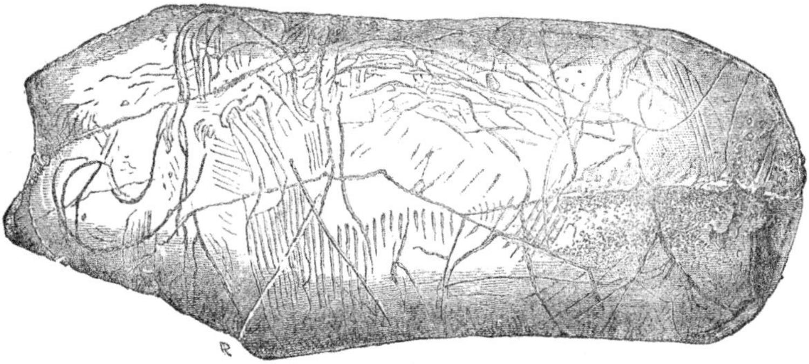

FIG. 108.—Figure of the Mammoth, engraved on Mammoth ivory by cave-men of La Madelaine, France.

as now flourishes in that temperate region. In the valley of the Tiber (where also its remains are found), though during the "Glacial period" the temperature was, doubtless, lower than at present, we cannot imagine that an arctic climate prevailed. Thus we see that it was capable of flourishing in various and widely separated regions where the conditions of climate and food supply could hardly have been similar.

Professor Boyd Dawkins, whose views we are adopting here,[1] considers that the Mammoth was exterminated by man—a simple

[1] *Popular Science Review*, vol. vii. p. 275 (1868).

solution of the question, which seems to present no difficulties. That it was hunted by the primitive folk of "the Reindeer period" in France, is proved by its remains in the caves where men dwelt, and by a drawing cut by a hunter of the Older Stone Age on one of its own tusks (see Fig. 108). A cast of this most interesting relic may be seen in the pre-historic galleries at the British Museum, Bloomsbury.

CHAPTER XV

GIANT SLOTHS AND ARMADILLOS

"Injecta monstris terra dolet suis."
HORACE, *Odes*, book iii.

IT would have been strange, considering how much we owe to North America, had the great South American continent not enriched our knowledge of past forms of life on the globe. But such is not the case. The honours are, as it were, divided, although it must be admitted that the North American extinct forms at present known are far more numerous. There are, however, two or three "Extinct Monsters" of very great interest which once had a home in South America—in that strange region of the Pampas, where the naturalist of the present day finds so much to excite his interest. Of these the present chapter treats.

The Megatherium [1] (Cuvier) was a gigantic mammal allied to sloths and ant-eaters, and perhaps to the armadillos. In its skull and teeth this colossus of the past resembled the sloths, in its limbs and backbone it resembled the ant-eaters, while in size it surpassed the largest rhinoceros (Plate XLI.). The famous, but imperfect, specimen at Madrid was for a long time the principal if not the only source of information with regard to this extinct genus, and for nearly a century it remained unique.

Later on, however, the zeal and energy of Sir Woodbine Parish, his late Majesty's *chargé-d'affaires* at Buenos Ayres,

[1] Greek—*megas*, great; *therion*, beast.

greatly helped to augment the materials for arriving at a just conclusion with regard to its proper place in the animal kingdom. According to one writer, Spain formerly possessed considerable parts of three different skeletons. The first and most complete is that which is preserved in the royal cabinet at Madrid. This was sent over in 1789, by the Marquis of Loreto, Viceroy of Buenos Ayres, with a notice stating that it was found on the banks of the river Luxan. In 1795 a second specimen arrived from Lima, and other portions, probably not very considerable, were in the possession of Father Fernando Scio, to whom they had been presented by a lady from Paraguay. But two German doctors, Messrs. Pander and D'Alton, who published in 1821 a beautiful monograph on the subject, state that they were unable in 1818 to find any traces of either the Lima specimen or that which had belonged to Fernando Scio.

The remains collected by Sir Woodbine Parish were discovered in the river Salado, which runs through the flat alluvial plains (Pampas) to the south of the city of Buenos Ayres, after a succession of three unusually dry seasons, "which lowered the waters in an extraordinary degree, and exposed parts of the pelvis to view as it stood upright in the bottom of the river."[1]

This and other parts having been carried to Buenos Ayres by the country people, were placed at the disposal of Sir Woodbine Parish by Don Hilario Sosa, the owner of the property on which the bones were found. A further inquiry was instituted by Sir Woodbine; and on his application, the governor granted assistance, the result of which was the discovery of the remains of two other skeletons on his Excellency's properties, at no great

[1] "Some Account of the Remains of the *Megatherium* sent to England from Buenos Ayres, by Woodbine Parish, Jun., Esq., F.R.S.," by Wm. Clift, Esq., F.R.S., *Geological Transactions*, second series, vol. iii., p. 437.

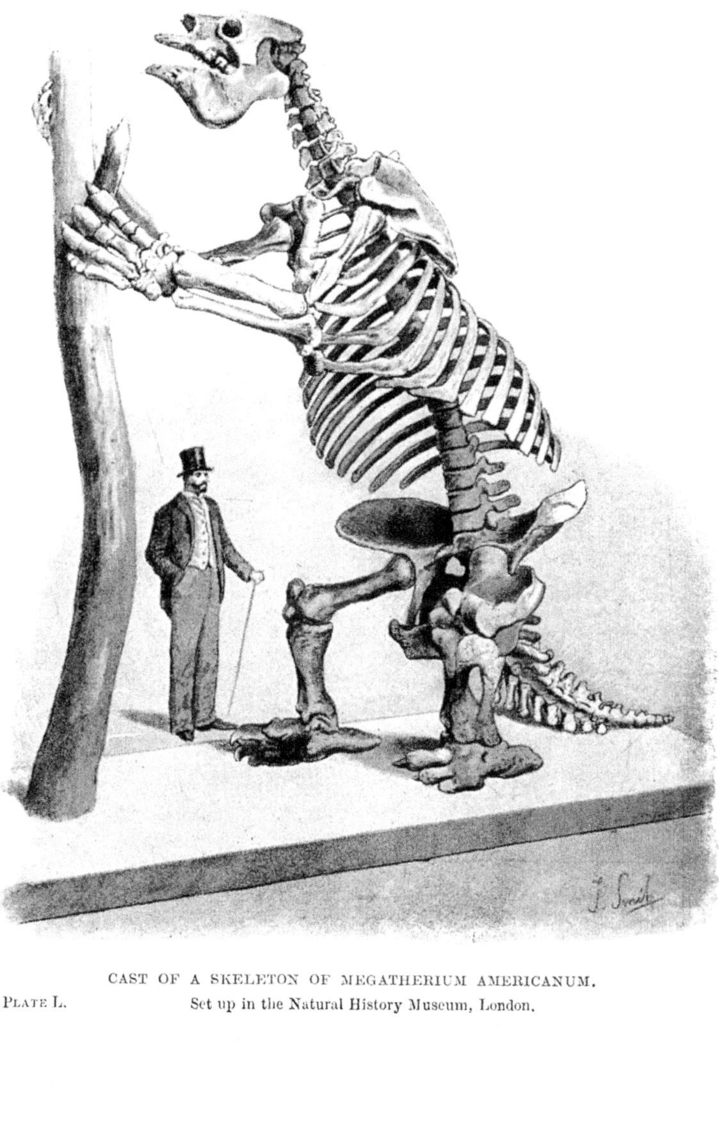

CAST OF A SKELETON OF MEGATHERIUM AMERICANUM.

PLATE L. Set up in the Natural History Museum, London.

distance from the place where the first had been found. It was in the year 1832 that Sir Woodbine Parish sent his valuable collection of bones from Buenos Ayres, and presented them to the Royal College of Surgeons. These specimens formed the subject of Mr. Clift's memoir above quoted. But even then the materials were not complete for a thorough knowledge of the bony framework of the Megatherium, and it was not till 1845, when more remains (discovered near Luxan, 1837) reached this country, that Sir Richard Owen was able to clear up one or two doubtful details. These were purchased by the trustees of the British Museum, and casts of the bones were taken. Among the various writings by learned men on the subject, Owen's masterly description stands pre-eminent; indeed, he was the only one to solve the riddle, to thoroughly explain the structure of this giant sloth, and to show how its food was obtained.[1] Neither Cuvier, nor the German doctors, nor Mr. Clift had succeeded in so doing.

In the Natural History Museum is a cast representing the animal nearly erect, and grasping a tree. This magnificent cast (see Plate L.) represents an animal eighteen feet in length, and its bones are more massive than those of the elephant. For instance, the thigh-bone is nearly thrice the thickness of the same bone in the largest of existing elephants, the circumference being equal to the entire length. To a comparative anatomist several striking indications of great strength present themselves; thus, not only the very forms of the bones themselves mean strength, but their surfaces, ridges, and crests are everywhere made rough for the firm attachment of powerful muscles and tendons. In the fore part of the body the skeleton is *comparatively* slender,

[1] His views are expounded in his *Memoir on the Megatherium, or Giant Ground Sloth of America*, 1861, which is beautifully illustrated. The Royal Society gave £100 (part of a Government grant of £1000) to enable Professor Owen to carry out this important work.

but the hind quarters show enormous strength and weight combined. The tail, also, is very powerful and massive. The fore limbs are long, and evidently constructed for the exertion of great force. How this force was applied we shall see presently. In both sets of limbs we notice powerful claws, such as might be used for scratching up the ground near the roots of a tree, and it was at one time thought that this was the way in which the creature obtained its leafy food, namely, by digging up trees by the roots and then devouring the leaves. But Owen had another explanation.

As in the living sloths and armadillos (edentata[1]), there are no teeth in the fore part of the jaw. The molar teeth, of which there are five on each side of the upper jaw, and four in the lower, are hollow prismatic cylinders, straight, seven to nine inches long, and implanted in deep sockets.

On looking at the model so skilfully set up at South Kensington, and especially at the front part of the skull, it will be seen that the snout and lips must have been somewhat elongated, possibly into a slight proboscis like that of the tapir. The specimens of the lower jaw in the wall-case close by show that it was much prolonged and grooved. This fact must be interpreted to mean that the creature possessed a long and powerful tongue, aided by which it could, like the giraffe, strip off the small branches of the trees which it had broken or bent down within its reach.

A bony shield (or carapace) of a great armadillo was found with one of the specimens described by Mr. Clift, and Buckland and others thought it belonged to the Megatherium; but Owen afterwards showed, by most clear and convincing reasoning from

[1] This word, which means *toothless*, is misleading. All the Pleistocene and existing edentata, however, agree in having no front, or incisor, teeth. Some forms from Santa Cruz have front teeth.

PLATE LI.

GREAT GROUND SLOTH OF SOUTH AMERICA, MEGATHERIUM AMERICANUM.
Length 18 feet.

the skeleton, that the Megatherium could not have been protected as armadillos are by such a shield (see p. 291).

And now we come to the question how it obtained its food. The idea of digging round trees with its claws in order to uproot them, must be partly, if not entirely, given up; for Owen has proved, by a masterly piece of reasoning, that this cumbrous creature, instead of climbing up trees as modern sloths do, actually pulled down the tree bodily, or broke it short off above the ground by a *tour de force*, and, in order to do so, sat upon its huge haunches and tail as on a tripod, while it grasped the trunk in its long powerful arms! Marvellous as this may seem, it can be shown that every detail in the skeleton agrees with the idea. Of course there would be limits to possibilities in this direction, and the larger trees of the period must have been proof against such Samson-like attempts on the part of the Megatherium; but when the trunk was too big, doubtless it pulled down some of the lower branches. Plate LI. is a restoration, by our artist, of the South Kensington skeleton.

Sir Woodbine Parish thought that the Megatherium fed on the Agave, or American aloe.

Another form of extinct sloth found in the same region is the Mylodon. Though of smaller size, it was much bigger than any living sloth, and attained a length of eleven feet. It has the same general structure, but the head and jaws are somewhat different, and more like the recent forms. A nearly perfect and original skeleton of Mylodon gracilis has been set up beside its huge relative's cast in the same gallery at the Natural History Museum. The crowns of its molar teeth are flat instead of being ridged; hence its name, which signified "mill-toothed."

Yet another was the Scelidotherium[1] with its long limbs.

[1] Greek—*scelis*, limb; *therion*, beast.

Darwin obtained an almost entire skeleton of one of these. It was as large as a polar bear. Speaking of his discovery, he says, "The beds containing the fossil skeletons consist of stratified gravel and reddish mud; a proof that the elevation of the land has been inconsiderable since the great quadrupeds wandered over the surrounding plains, and the external features of the country were then very nearly the same as now. The number of the remains of these quadrupeds embedded in the vast estuary deposits which form the Pampas and cover the granitic rocks of Banda Oriental must be extraordinarily great. I believe a straight line drawn in any direction through the country would cut through some skeleton or bones. As far as I am aware, not one of these animals perished, as was formally supposed, in the marshes or muddy river-beds of the present land, but their bones have been exposed by the streams intersecting the subaqueous deposit in which they were originally embedded. We may conclude that the whole area of the Pampas is one wide sepulchre of these extinct gigantic quadrupeds."[1]

The genus Scelidotherium comprises a number of species and presents characters more or less intermediate between Megatherium and some other genera. The skull is low and elongated, and shows an approach to that of the modern ant-eater. The feet also are different from those of Megatherium (see Fig. 109).

These monster sloths inhabited South America during the latest geological period, known as the Pleistocene. During part of that time North America, as well as Northern Europe and Asia, were invaded by a great ice-sheet, and an arctic climate prevailed. It is therefore very probable that while the mammoth and the mastodon were roaming over North America, giant sloths and armadillos were monarchs of the southern continent. What

[1] *Journal of Researches.*

cause, or causes, led to the extermination of the giant sloths and armadillos is still a matter of speculation. One writer suggests an explanation that seems to deserve consideration. The southern parts of this great continent are even now subject to long-continued droughts, sometimes lasting for three years in succession, and bringing great destruction to cattle. In fact, the discoveries related above were rendered possible by several successive dry seasons. It is argued that the upright position of most of the skeletons found *in situ* seems to suggest that the creatures must have been mired in adhesive mud sufficiently firm

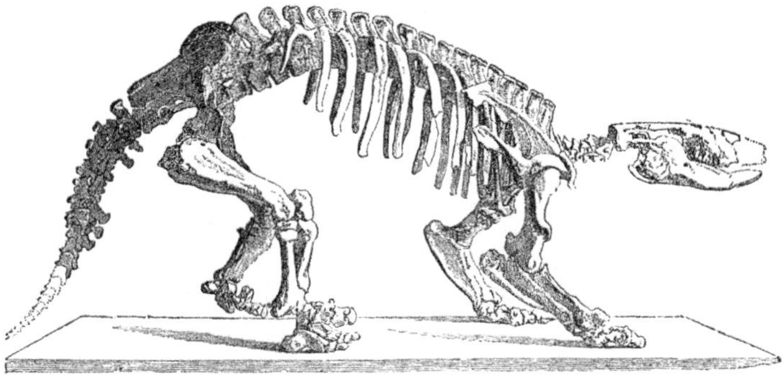

FIG. 109.—Skeleton of *Scelidotherium*. (After Capellini.)

to uphold the ponderous bones after the flesh had decayed. A long drought would bring the creatures from the drained and parched country to the rivers, reduced by want of rain to slender streams running between extensive mud-banks; and it is possible that, in their anxious efforts to reach the water, they may have only sunk deeper and deeper in the mud until they were engulfed. This idea is strengthened by information supplied to Mr. Darwin when in these parts (recorded in his *Journal*). An eye-witness told him that during the *gran seco*, or great drought, the cattle in herds of thousands rushed into the Parana, and, being exhausted by hunger and thirst, were unable to crawl up the muddy banks, and so were drowned.

In the last great drought, from 1830 to 1832, it is probable (according to calculations made) that the number of animals that died was over one million and a half. The borders of all the lakes and streamlets in the province were long afterwards white with their bones.

It does not seem probable that the small sloths of to-day are descended from these big ones; but it may be that some of the arboreal forms forsook their trees and took to the ground. But nobody can say why they became so enormous.

There can now be no doubt that in the southern part of South America the giant sloths survived until the human period; for in the Argentine Republic their remains have been found together with human bones and stone weapons. Dr. A. Smith Woodward, the Keeper of Geology in the Natural History Museum, thus describes this highly interesting find. "The most important discoveries, however, which appear to prove this survival, were made in 1897, and subsequent years, by Dr. F. P. Moreno, Dr. R. Hauthal, Baron Erland Nordenskiöld, and others, in a cavern near Consuelo Cove, Last Hope Inlet, Patagonia, between the 51st and 52nd degrees of south latitude. Here, in an absolutely dry and powdery deposit on the floor of the large cavern, were found numerous broken bones of several individuals of a ground-sloth, Grypotherium, which was nearly as large as Mylodon, and only differed from the latter in minor features. With the bones were several pieces of skin, evidently of the same animal, which showed marks of tools, and seemed to have been stripped off the carcase by man. There were also large lumps of excrement, besides masses of cut grass, which may have been intended for fodder. With the Grypotherium were found

THE LATE SIR RICHARD OWEN.

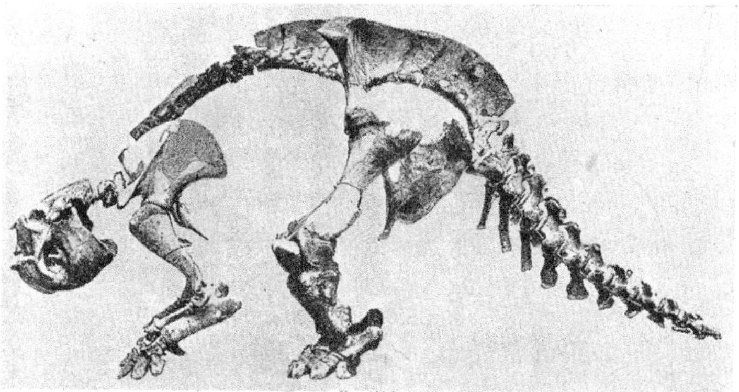

SKELETON OF GLYPTODON CLAVIPES (IMMATURE, WITHOUT CARAPACE).

From the Pampas formation, Buenos Aires, La Plata Museum. (After Lydekker.)

PLATE LII.

bones of other extinct animals; and in the same cavern were implements of stone and bone, remains of fires, and even the bones of man himself. The Argentine explorers, in fact, concluded that the Grypotherium had actually been kept in the cavern and fed by man, who eventually killed the animals for food." At the Natural History Museum may be seen a series of specimens illustrating this discovery. The pieces of skin are covered with coarse hair on the outside, and on the inside are small nodules of bone. These bony nodules (found with other skeletons in the Pampas formation) are of special interest: first, because their presence had never before been even suspected, and secondly, because they serve to some extent to strengthen the bonds of relationship between Sloths and Armadillos, which have a heavy armour (see p. 292). An extinct horse was also found in this cave.

We now pass on to the old gigantic representative of the armadillo, the Glyptodon.[1] To the eye it resembles more or less an armadillo, and has a huge cuirass, or large plate of armour, covering the whole of the body, but allowing the head to show in front, while the legs come out beneath. Both head and tail were also protected with armour. The great shield, or carapace, in most of the extinct armadillos, is composed of bony plates of regular shape, closely united at their edges (sutures) so as to form a solid piece. It is evident, therefore, that this creature, having no movable bands, as living armadillos have, could not roll itself up into a ball. The fore feet have thick, short toes, instead of long ones, such as their modern representatives have; and from this we may infer that they were not in the habit of burrowing or of seeking their food underground. The family of

[1] So named by Sir R. Owen, in reference to the sculptured aspect of the grinding surface of the teeth. Greek—*glupho*, I carve; *odous*, *odontos*, tooth.

Glyptodonts seem to have been chiefly confined to the continent of South America, but some species are known to have extended their range as far as Mexico, and Texas into North America. A good deal of confusion has arisen with regard to the classification of these old-fashioned armadillos, on account of the fact that isolated specimens of their tails have often been found, and these cannot always be referred to the right carapaces. For example, it should be pointed out here that the tail represented in Plate LIII. really belongs to another genus, known as Hoplophorus.[1]

In Glyptodon asper the scutes of the carapace had a beautiful rosette-like sculpture, while the sheath of the tail was entirely composed of a series of movable rings, ornamented with large projecting tubercles. The vertebræ of the backbone are almost entirely fused together into a long tube, and also are joined to the under surface of the great shield, to which the ribs are united. The cheek-teeth are sixteen in number, four above and four below on each side. These are channelled with two broad and deep grooves, which divide the surface into three distinct lobes. Hence the name of the animal.

Plate LII. represents Glyptodon clavipes (Owen) from the Pleistocene deposits of Buenos Ayres; but the reader will gain a much better idea of the animal by inspecting the splendid specimen of Glyptodon asper in the Natural History Museum, near the centre window at the East end of the Gallery of fossil mammals.

Plate LIII. is a restoration of another species by our artist.[2]

In the Museum of the Royal College of Surgeons (which the reader is recommended to visit) there are several most valuable specimens of these extinct armadillos from South America.

[1] Greek—*Hoplon*, armour; *phero*, I bear.

[2] This plate is based on a beautiful drawing in a Spanish work, *Anales del Museo publico de Buenos Aires.* G. Burmeister, M.D., Phil. D. Tomo Segundo.

A GIGANTIC ARMADILLO, GLYPTODON ASPER. PLEISTOCENE PERIOD. SOUTH AMERICA.

PLATE LIII.

From Buenos Aires. Length 8 feet 7 inches.

Armadillos belong, with sloths and ant-eaters, to the same family of so-called toothless animals (edentata) with no front teeth, though one or two forms really are toothless. Those of the present day have their bony armour divided up into a series of bands, so that they can roll themselves up, more or less, into balls. They burrow under the ground, where they get their food to a certain extent, and live a safe life, protected by their casque of mail. Their only enemies seem to be the monkeys, and one of the tricks of the young monkeys in the American forests is, when they find an armadillo away from home, to pull its tail unmercifully, and try to drag it about. Snakes cannot hurt them. Mr. W. H. Hudson, in his most interesting book, *A Naturalist in La Plata,* narrates how he watched an armadillo kill a snake and then devour it.

If we examine the anatomy of the armadillo, we shall find that its bones greatly resemble those of the sloth, but still there are a few differences. It is a burrowing animal, and therefore requires great power of scratching and tearing the ground. Why the colossal forms of armadillo should have become extinct and only small ones survived to the present time, is one of the many and perplexing problems presented by the study of extinct animals. One would have thought from its size and strength that the Glyptodon had been built, like Rome, for eternity.[1]

[1] The reader who is interested in evolution is strongly recommended to read the very interesting and suggestive address of Dr. A. Smith Woodward, F.R.S., to the Geological section of the British Association, 1909 (Winnipeg Meeting). In this paper the writer speaks of the symptoms of old age in races. One of these is loss of teeth, shown in tortoises, birds, monotreme mammals of Australia, and in the Ichthyosaurs. Another is degeneration into eel-like forms, and another is a superfluity of dead matter in the shape of spines or bosses, armour plate, and big horns, like those of the Irish deer (Cervus giganteus).

CHAPTER XVI

EXTINCT WHALES AND WOMBATS

"The earth from her deep foundations unites with the celestial orbs that roll through boundless space, to declare the glory and show forth the praise of their common Author and Preserver; and the voice of Natural Religion accords harmoniously with the testimonies of Revelation, in ascribing the origin of the Universe to the will of One eternal and dominant Intelligence, the Almighty Lord and supreme First Cause of all things that subsist, 'the same yesterday, to-day, and for ever.'"—DEAN BUCKLAND.

IT would not be safe to venture any positive statement or opinion as to when the great cetacean order of mammals, which includes such creatures as whales, dolphins, and porpoises, first appeared on the earth. But as far as the geological record is known at present, it would seem as if they were ushered in at the commencement of the Tertiary era, at the same time with many other new forms of life. The original ancestor of all these marine mammals may, for all we know, have been evolved some time before then; say in the Cretaceous period, or perhaps during that great and partly unknown interval between the Secondary and Tertiary eras. What we do know for certain is, that fossil evidence of their existence first comes to light in the Eocene strata of North America.

The oldest known whale is the curious Zeuglodon, from the marine Eocene strata of Alabama. At Claiborne, according to Sir Charles Lyell, there occur numerous species of shells, besides many sea-urchins (echinoderms) and abundance of sharks' teeth. But the most remarkable remains found there are those of the extinct whale, known as Zeuglodon cetoides,[1] and so called from the yoke-like form of its double-fanged molar teeth, which are six inches in length. It has also been found in Mississippi, Georgia, South Carolina, and Egypt. No strata of later date than the Eocene period contain its remains, as far as we know; and it has never been found out of the Northern Hemisphere. Zeuglodon must certainly have been a very remarkable creature, and it is to be hoped that palæontologists will some day meet with a complete skeleton, such as may afford the material for showing its entire bony structure. Unfortunately, although a great many finds have been made in Alabama, no complete skeleton has yet been got together. The remains of at least forty individuals have been met with, so that we may venture to say that this ancient cetacean once flourished vigorously. So numerous are the vertebræ in the above-mentioned locality that it is said they are used for making fences, and that farmers even burn the bones in order to get rid of them! How little they know that hundreds of museum curators all over the world would be only too glad to procure some of this "rubbish"! With the spread of education, and an increasing desire to learn a little about the former history of our world, let us hope that in time agriculturists will endeavour to co-operate with men of science in rescuing from destruction the fossilised remains of the earth's former inhabitants.

Sir R. Owen declared the Zeuglodon to have been "one of the most extraordinary of the Mammalia which the revolutions of the globe have blotted out of the number of existing beings." Some of the vertebræ found in Alabama have a length of eighteen inches, and a diameter of twelve inches, while the whole of the backbone, or vertebral column, has been found undisturbed.

[1] Greek—*zeuglé*, yoke; *odous*, *odontos*, tooth.

Sir Charles Lyell said: "The vertebral column of one skeleton found by Dr. Buckley at a spot visited by me, extended to a length of nearly seventy feet, and, not far off, part of another backbone, nearly fifty feet long, was dug up. I obtained evidence, during a short excursion, of so many localities of this fossil animal within a distance of ten miles, as to lead me to conclude that they must have belonged to at least forty individuals."

This animal is particularly interesting as marking the first appearance of the very distinct order of cetaceans, so unlike any other order, except the Sirenia (Manatee and Dugong).

Professor D'Arcy Thompson thinks that Zeuglodon and other members of its family are related to the seals, and have no direct affinities with the cetacea; their teeth certainly are very similar, but unfortunately we do not at present know what their limbs were like—though one would expect, from the long pointed head of Zeuglodon, that the body also was long like a whale's, and adapted for rapid progress through the water, rather than for progression on land. One cannot easily picture a creature with such a skull walking on the land, even in the awkward manner of a seal; but we must wait for further evidence on this point. A portion of a skull of Zeuglodon was found in the Barton Clay, an Eocene deposit in Hampshire.[1]

We have already recorded one or two cases in which curious mistakes have been made by palæontologists; but the student soon finds how many and great are the difficulties that have to be overcome by those who study lost forms of life, difficulties which in most cases are due to the imperfection of the material

[1] The late Professor Dames, of Berlin, has described certain bony plates found with Zeuglodon remains, which probably indicate that its back was protected by an armour of scutes (see *Natural Science*, vol. iv. p. 175).

at their command, or the bad state of preservation in which the specimens are often found. The old saying, "Humanum est errare," is particularly true of palæontological workers. But, at least, it can be said of them that they are honest, and never attempt to delude the public. There have, however, occasionally arisen enterprising persons, more desirous of making money than of furthering scientific truth, who have, for a time, succeeded in the grossest deceptions. Matzuyer, about 1613, exhibited some Mastodon bones as those of a human giant (Teutoboccus rex) and made a great stir in France; and now we have to record another case in connection with Zeuglodon.

In the year 1845, Dr. Albert Koch exhibited a large skeleton of a fossil animal, under the name Hydrarchos Sillimani, the former of which would mean "King of the Sea," while the latter seemed to indicate the approval of Professor Silliman. (The name is tempting to a punster.) These remains consisted of a head and vertebral column, measuring in all one hundred and fourteen feet, a few ribs and parts of some supposed paddles. Dr. Koch made the public believe that all the bones had been found together in such a way as to prove that they belonged to a single individual, and that the numerous vertebræ formed a connected series. This, however, was a mere fabrication, as was shown by Professor Wyman, who says "these remains never belonged to one and the same individual," and that the anatomical characters of the teeth prove them to be, not those of a reptile, but of a warm-blooded mammal. He came to the conclusion that the greater part of the bones belonged to the genus Basilosaurus (King-saurian) of Harlan, a name which the American geologists still retain for Zeuglodon. This skeleton represented *at least* two separate individuals.

The following letter on the subject, from the well-known

geologist, Dr. G. A. Mantell, appeared in *The Illustrated London News* of November 4, 1848:—

"SIR,

"Will you allow me to correct a statement that appeared in the last number of your interesting publication? The fossil mentioned at the conclusion of the admirable notice of the so-called sea-serpent, as having been exhibited in America under the name Hydrarchos Sillimani, was constructed by the exhibitor, Koch, from bones collected in various parts of Alabama, and which belonged to several individual skeletons of an extinct marine cetacean, termed Basilosaurus by the American naturalists, and better known in this country by that of Zeuglodon, a term signifying yoked teeth. Mr. Koch is the person who, a few years ago, had a fine collection of fossil bones of elephants and mastodons, out of which he made up an enormous skeleton and exhibited it in the Egyptian Hall, Piccadilly, under the name Missourium. This collection was purchased by the trustees of the British Museum, and from it were selected the bones which now constitute the matchless skeleton of a mastodon in our National Gallery of organic remains [then in the British Museum]. Not content with the interest which the fossils he collected in various parts of the United States possess, Mr. Koch, with the view of exciting the curiosity of the ignorant multitude, strung together all the vertebræ he could obtain of Basilosaurus, and arranged them in a serpentine form, manufactured a skull and claws, and exhibited the monster as a fossil sea-serpent, under the above-mentioned name, Hydrarchos. But the trick was immediately exposed by the American naturalists, and the true nature of the fossil bones pointed out.[1]

[1] After this, Koch's sea-serpent was carried to Dresden.

"Bones of the Basilosaurus have been found in many parts of Alabama and South Carolina, in greensand belonging to a very ancient Tertiary formation; hundreds of vertebræ, bones of the extremities, portions of the cranium and of the jaws with teeth have from time to time been collected. Remains of species of the same genus have also been found near Bordeaux and in Malta.

"Professor Owen has shown that the original animal was a marine cetacean, holding an intermediate position between the Cachalots and the herbivorous species. It must have attained a length equal to that of the largest living whales, for a series

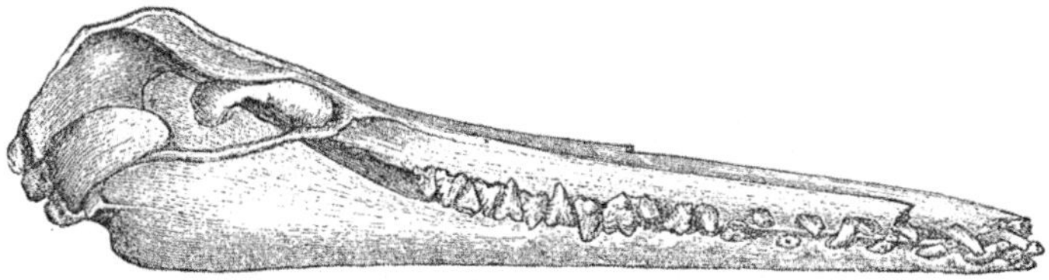

FIG. 110.—Head of *Squalodon*, from Miocene strata, Bavaria.

of vertebræ was observed *in situ* that extended in a line sixty-five feet. . . .

"GIDEON ALGERNON MANTELL.

"19, Chester Square, Pimlico, Oct. 31, 1848."

Squalodon (see Fig. 110) represents another extinct genus of cetaceans, of which very little is known beyond the teeth and skull. It was formerly classified with Zeuglodon, but its skull is more like that of a dolphin, although the teeth are somewhat like those of the former. Teeth and fragments of skulls of Squalodon have been frequently found in the marine European deposits of Miocene age—especially in the Vienna basin, many parts of France, and the Crag formations of Antwerp and of Suffolk; also in strata of corresponding age in North America and New Zealand. A few isolated teeth occur in the cave-deposits of Italy. The Natural History Museum contains an excellent reproduction of the skull of Squalodon.

The researches of Dr. C. W. Andrews (see his catalogue referred to on p. 266) and Dr. E. Fraas in Egypt have revealed several earlier ancestors of the whales. Of these Protocetus appears to be derived from the Creodonts (early ancestors of the Carnivora). This is followed by Prozeuglodon.

Remains of the gigantic Sperm Whale or Cachalot (Physeter macrocephalus) are found in the Forest-bed of Cromer, and also

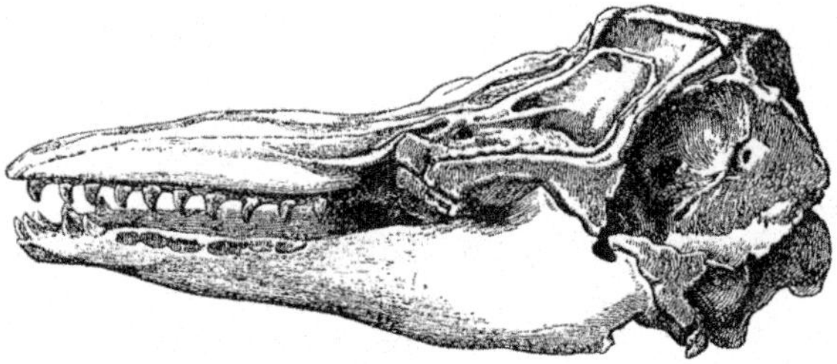

FIG. 111.—Skull of thick-toothed Grampus, *Pseudorca crassidens*. (Owen.) Fen, Lincolnshire.

in a later deposit in South America. A number of Miocene and Pliocene forms allied to the Cachalot have been described under various names. Remains of the curious Narwhal, with its long spear-like tusk, have been found in the Norfolk forest-bed. Owen has described the skull of a thick-toothed Grampus that was found beneath the turf in the great fen of Lincolnshire, in the year 1843 (see Fig. 111). It belongs to the existing Pseudorca crassidens, but Owen thought it differed from the present dolphins of our own coasts. Extinct genera of dolphins are known to occur in some of the later Tertiary deposits of Europe and America.

That whales were contemporary with the mammoth in this country is proved by an interesting discovery made in the year 1828, in a cliff near Brighton, under the following circumstances. On the face of the cliff, in the ancient shingle which lies immediately upon the chalk, some fishermen had observed a huge bone that had been laid bare by an unusually high tide, and now projected two or three feet beyond the face of the cliff. Unable to remove it, they broke off the extremity, and sent a fragment of it to Dr. Mantell, who went to the spot a few days afterwards, and found that they had demolished a considerable portion of the bone in their attempts to remove it from its bed, and had only desisted from fear of being buried beneath the overhanging cliff, which is composed of loose materials. At last, and not without some danger, the entire specimen was left exposed. It proved to be the front nine feet of the left branch of the lower jaw of a whale-bone whale. On attempting to remove the specimen it broke into a thousand pieces. So, as the tide was approaching, this interesting find had to be abandoned. It must have belonged to a whale from sixty to seventy feet long.

Some of the ancient kangaroos of Australia, that lived during the Pleistocene period, were very much larger than those we see now in that country. One of the most remarkable of the extinct marsupials of Australia was the Diprotodon, of which a skull, measuring nearly three feet in length, may be seen in the Natural History Museum, and it was so named on account of its two front teeth. The creature itself was probably even larger than a rhinoceros of the present day (standing six feet high), and is believed to have resembled a modern Wombat. The restored skeleton shown in Plate LV. will be presently explained. The history of the gradual discovery of this strange marsupial may now be briefly given, since it serves to show how little ground there is for the popular belief that a Cuvier or an Owen could "restore" a whole animal from a jaw, or even a single tooth! Sir R. Owen himself was obliged to admit that his first conclusion, drawn from

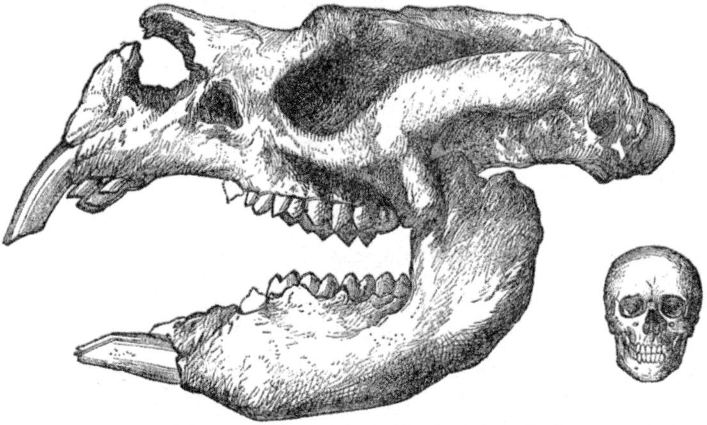

FIG. 112.—Skull of *Diprotodon*, a kind of Wombat, from Pleistocene strata, Australia. A man's skull placed for comparison.

part of a jaw of Diprotodon, was quite wrong, as we shall see from the following account.

The name[1] was given by Owen to part of a lower jaw of the creature, obtained by Sir Thomas Mitchell in the Wellington Caves, New South Wales. Five years later, a drawing of part of a jaw with teeth reached England from the same source, and this Sir R. Owen believed to represent a kind of Dinotherium, indicating, for the first time, the occurrence of primitive elephants in Australia! In the same year, a portion of a molar tooth, associated

[1] Greek—*dis*, twice; *protos*, first; *odous*, tooth.

with part of a thigh-bone and other fragments, was also received, and Owen wrote: "The fossils which my friend has now transmitted incontestably establish the former existence of a huge proboscidian pachyderm in the Australian continent, referable either to the genus *Mastodon* or *Dinotherium* (!)."

Only a year later, however (1844), this prophecy proved to be incorrect; and, within a short time, Sir R. Owen was able, not merely to describe correctly most of the Diprotodon's skeleton, but also to distinguish another allied genius, the Nototherium. The feet alone remained unknown, and part of these were described as the toes of that fabulous monster, the "Great horned Lizard of Australia"!

In 1863 our late distinguished friend, Dr. George Bennett, wrote from Sydney to Owen, saying that his son, then in Queensland, had written home to say that he found a place where a whole skeleton lay: the bones were immense; the head, he said, had been sent to Sydney some years previously (this may be the one now in the Natural History Museum). Owen wrote to Sir Henry Parkes, in 1867, suggesting that his administration should send an exploring party to the limestone caves in Wellington Valley, discovered in 1832, and from which fossil remains were brought sufficient to show that the marsupial type of mammal formerly prevailed in that country as it does now. Consequently, a vote of two hundred pounds was passed, and a most valuable collection of fossils obtained.

The following passage, from a now historic paper by Sir R. Owen, will serve to show with how great enthusiasm he pursued his most fruitful researches in palæontology. "Of no existing animal," he says, "of which a passing glimpse, as it were, had thus been caught, did I ever feel more eager to acquire fuller knowledge than of this huge Marsupial. No chase can equal the excitement of that in which, bit by bit, and year after year, one captures the elements for reconstructing the entire creature of which a single tooth or fragmentary bone may have initiated the quest; in the course of which one fully realises, with more or less exactitude, the picture which the laws of correlation had led one to frame of an animal which may have passed out of existence long ages ago."[1] Owen's restoration, published in the

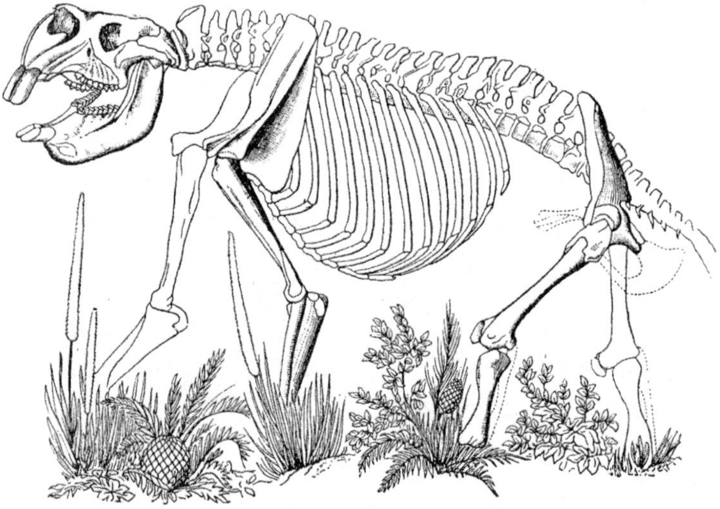

Fig. 113.—Skeleton of a huge extinct marsupial, *Diprotodon australis*. (Restored after Owen.)

above paper, is shown in Fig. 113. But, in the year 1893, a wonderful find was reported from Australia, which will, doubtless, in time lead palæontologists to a complete and thorough knowledge of this great and mysterious marsupial. The first announcement was a brief telegram in *The Times;* but, later on, a letter appeared in *The Scotsman,* of September 14th, from Mr. William Kinmont. In the same enterprising newspaper

[1] *Philosophical Transactions of the Royal Society*, vol. 160 (1870), p. 519.

appeared a few days afterwards (September 25th), extracts from an Australian account of the matter, the *South Australian Register* of August 12th. It seems that only a small portion of the area known to contain fossil remains has as yet been explored, and there is little doubt that a complete examination will yield results even surpassing those already made.

Mr. Kinmont wrote as follows:—

"Birksgate, Glenosmond, Adelaide, S.A., August 9, 1893.

"SIR,

"Some two months ago a discovery of mammoth fossil bones was made at Lake Mulligan in South Australia, which promises to be of incalculable value to science, and the 'record' discovery of the kind for the century. As the subject is one of world-wide interest, I propose to give you in brief, from my own knowledge and observation, the facts and particulars of the discovery.

"Lake Mulligan is situated in the central district of South Australia, and a more apparently uninteresting and forsaken spot than this locality could not be found in the colony. When the famous Australian explorers, Captain Sturt and Mr. John M'Douall Stuart, went north-west from Menindie on the river Darling, in 1844, they came into a great stony desert lying to the eastward of Lake Frome, and the party endured the most terrible hardships from the frightfully barren nature of the country, and the extraordinary heat of the weather. . . .

"What a different picture of the past history of this country is brought to light by the recent discoveries! On the sides of these mountains lying between Lake Frome and Lake Torrens must have been grown huge trees, and all around there must have been a dense tropical growth, exceeding in luxuriance the forests of the eastern slopes of the Andes in South America. Every form of life was at that period huge and uncouth. On the grassy plains at the foot of the mountains roamed the Diprotodon, a gigantic animal allied to the kangaroo in its structure and habits, but of the size of an elephant.

"But to return to the recent discovery of the actual bones. The South Australian Museum, represented by the honorary director, Dr. E. C. Stirling, C.M.G., F.R.S., took the matter in hand, and despatched a party of competent men to excavate. These men have now been at work for some four months removing the gravel that buries the bones, and the discovery has been proved to be even greater than was anticipated, and already many new discoveries of the first importance have been made. For the first time in the history of palæontological science the complete skeleton of Diprotodon australis has been made known—this being a huge marsupial considerably exceeding the rhinoceros in size. The bones of this magnificent treasure, in the interests of science, are now in the Adelaide Museum. Remains also of a giant wombat as large as a half-grown bullock, of two or three kinds of gigantic birds equalling the Moa of New Zealand in size, several species of colossal kangaroos, as well as the head of what is undoubtedly a new species of Diprotodon, have been discovered, excavated, and brought to Adelaide. Over seventy different extinct animals and birds, hitherto unknown, are represented in the two thousand bones which have been dug out of the valley at Lake Mulligan.

"There can, in fact, be no doubt but that this discovery is the most remarkable one of its kind that has ever been made in Australia, and perhaps in the world. The South Australian Museum, with its limited funds, has done what it could, but with the work only fairly commenced the directors found themselves in the position of having to abandon the search for lack of means to go on. Sir Thomas Elder, however, the distinguished initiator and promoter of Australian exploration expeditions, with his usual liberality to the cause of science, and with a desire to advance the scientific reputation of South Australian work, has come to the assistance of the museum in this matter, and has enabled the directors to continue this most important and extensive work. Dr. Stirling proposes to start for Lake Mulligan himself this week, with a view of personally superintending operations; and there can be little doubt but that work in the future will yield results even surpassing those already made.

"I am, etc.,

"WILLIAM KINMONT."

The following is part of the article above alluded to from the *South Australian Register*:—

"Sir Thomas Elder, G.C.M.G., has once more placed South Australia under deep obligations to him by coming forward with the monetary help necessary to enable further explorations to be made of the area at Lake Mulligan, in which immense quantities of fossil remains of extinct animals and birds are known to exist. The generous assistance thus rendered has come at the nick of time. The authorities of the South Australian Museum, recognising the importance of the discovery in the interests of science, have already done much to secure specimens of the bones which have been preserved from destruction; but the funds at their disposal are limited, and, unaided, they would have been unable to have carried on excavations except upon a very moderate scale. Owing to the munificence of Sir Thomas Elder, the search party, under charge of Mr. Hurst, which has succeeded in sending down a sufficient number of specimens to whet the appetite of scientists, and even of the curious public, who have no special knowledge of palæontology, for more, will now be reinforced by Dr. Stirling, Mr. Zietz, and another member of the Museum staff. More than that, provision will now be made for the conveyance to Adelaide of larger portions of the remains than it has hitherto been possible to despatch.

"Mr. Hurst has laboured under great disadvantages in the past, owing to the lack of facilities for transporting to the city the results of his excavations. It is to be borne in mind that the fossils are spread over a large tract of country embracing several square miles in extent. Moreover, this singular burial-place of the marsupial and other monsters, which were wont to roam over the continent of Australia, is situated in a remote and comparatively waterless region, so that the difficulties in the way of disinterment and removal are exceedingly great. It is only fitting that those officials of the Museum, who are most skilled in natural history, should be on the spot to superintend operations, and we may rest satisfied that everything that can be done to secure the best illustrations obtainable of the fossil remains will be done. The object to be aimed at is to procure complete specimens, and, from what has been said by those who have visited the locality, it should be practicable to achieve this. Obviously no pains or expense should be spared to bring together, bone to his bone, the framework of the gigantic creatures belonging to a bygone age, whose relics are embedded in the hardened mud and *débris* of what were once the huge swamps of the Far North-East."

The late Professor Newton, of Cambridge, also received a letter from Dr. Stirling, which was published in *The Proceedings of the Zoological Society* for May 2, 1893. From this letter we learn that Diprotodon has five well-developed toes; the bones of the

PLATE LIV. Fore limb. Hind limb.

THE REMAINS OF THE DIPROTODON AS FOUND AT LAKE CALLABONNA, S. AUSTRALIA.

marsupium, or pouch, have been found, and several distinct impressions of the skin of the fore foot were obtained. The locality is six hundred miles north of Adelaide, and, during the dry season, all the carting and travelling has to be done by camels, with a temperature of 111° Fahr. in the shade!—to say nothing of myriads of flies and frequent sandstorms.

With regard to the geological formation of this salt-lake district of South Central Australia, Lake Mulligan is a vast level expanse of salt-encrusted, black mud, only becoming filled after very heavy rains (see Plate LIV.). It is about eight miles wide, and the bones are found somewhere about midway between the east and west edges. Usually the salt crust is not firm enough for bullock traffic, and it is probable that thousands of bullocks have at different times been bogged in crossing or attempting to cross.

Several skeletons of a large wombat, about the size of a bullock, have been unearthed, probably Phascolomys gigas; also another slender creature about the size of a sheep, which is as yet an unknown animal. The meaning of this great find of bones would appear to be that immense herds of Diprotodon and other creatures, in seeking for water in a dry season, got bogged, just as cattle do now in the North, by hundreds. Crocodiles and alligators inhabited the fresh-water lakes that once covered parts of the district, either in the Pliocene or Pleistocene period—or both.

It now only remains to relate very briefly the progress made since the present writer published the above account in the year 1894. Palæontologists have for years anxiously awaited the reconstruction of Diprotodon by Dr. E. C. Stirling, Mr. Zietz, and their colleagues of the South Australian Museum, at Adelaide.

Since the year 1893 numerous complete skeletons have been found. However, very great difficulties have to be overcome in digging out these very fragile (and crushed) bones, and transporting them on the backs of camels; so that even now some parts of the skeleton are more or less unknown. As the result of years of labour, a "restoration" of the skeleton has been set up in the South Australian Museum, of which a plaster copy was presented to the Cambridge Museum of Zoology. The authorities of the former Museum have also been good enough to present to the Natural History Museum a set of actual limb-bones, and vertebræ of the tail, together with enough plaster casts to make the highly interesting restoration shown in Plate LV. In this specimen the head and the vertebræ from the neck to the sacral region (*i.e.* to where the hind limbs are situated) are exact copies of the Adelaide reconstruction. The skull, unfortunately, is not to be relied upon, being only a reconstructed model based on specimens from Lake Callabonna, which are all sadly crushed. Hence we are still in ignorance of the precise shape of the skull. A recent examination of the specimen from Queensland described by Owen shows that nearly all the brain-case is wanting, and that the skull was four or five inches shorter than Owen supposed. The two scapulæ, or shoulder blades, in the London model, are casts from two actual specimens from Queensland; and the feet are also casts from actual specimens. Possibly before long the whole model may be reconstructed. Meanwhile, palæontologists will await with a lively interest, the publication of a promised memoir by Dr. Stirling and Dr. Zietz. Mr. Dollo has suggested that possibly the ancestors of Diprotodon were arboreal creatures, a conclusion based on the evidence of the feet.

In conclusion, it only remains to express a hope that the reader may have been interested in our humble endeavours to describe

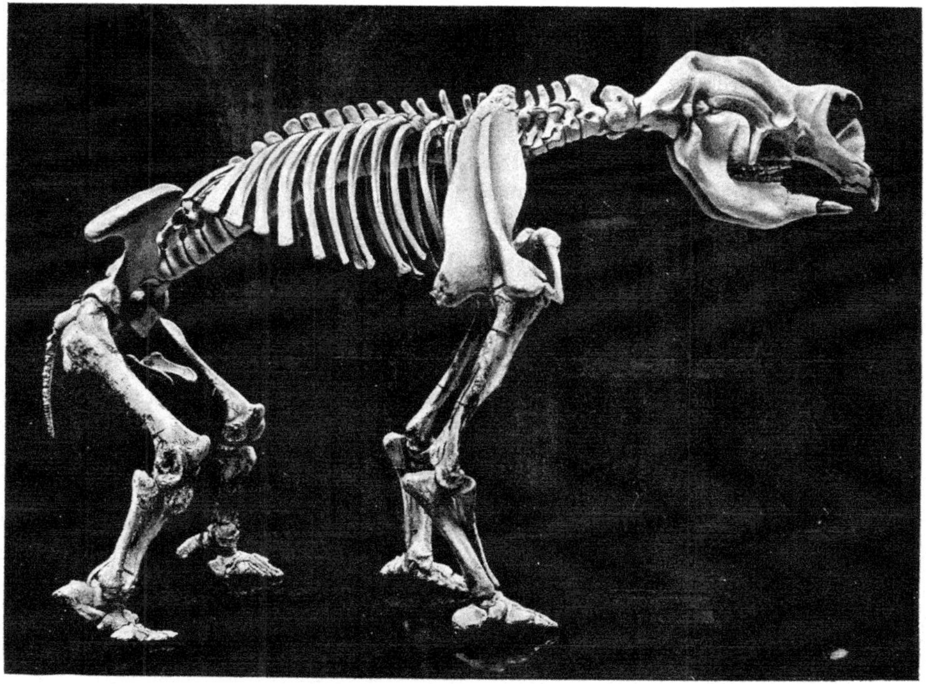

Plate LV.

RESTORED SKELETON OF DIPROTODON AUSTRALIS.
In British Museum (Natural History), London.

some of the largest, most strange, and wonderful forms of life that in remote ages have found a home on this planet. And perhaps a ew of our readers may be induced to add a new and never-failing interest to their lives by searching in the stony record for traces of the world's "lost creations." If so, our labour will not have been in vain.

APPENDIX I

Table of Stratified Rocks

Periods.		Systems.	Formations.	
CAINOZOIC.	Quaternary.	**RECENT** 200 ft.	Terrestrial, Alluvial, Estuarine, and Marine Beds of Historic, Iron, Bronze, and Neolithic Ages	Dominant type, Man
		PLEISTOCENE 450 ft.	Peat, Alluvium, Loess Valley Gravels, Brickearths Cave-deposits Raised Beaches Boulder Clay and Gravels	
	Tertiary.	**PLIOCENE** 1500 ft.	Norfolk Forest-bed Series Norwich and Red Crags Coralline Crag (Diestian)	Dominant types, Birds and Mammals
		MIOCENE 1000 ft.	Œningen Beds Freshwater, etc.	
		OLIGOCENE 600 ft.	Fluvio-marine Series	
		EOCENE 800 ft.	Bagshot Beds } (Nummulitic London Tertiaries } Beds)	
SECONDARY, OR MESOZOIC.		**CRETACEOUS** 2500 ft.	Maestricht Beds Chalk Upper Greensand Gault Lower Greensand } Neocomian Wealden }	Dominant type, Reptiles
		JURASSIC 5000 ft.	Purbeck Beds Portland Beds Kimmeridge Clay { (Solenhofen Beds) Corallian Beds Oxford Clay Great Oolite Series Inferior Oolite Series Lias	
		TRIASSIC 3000 ft.	Rhætic Beds Keuper Muschelkalk Bunter	

TABLE OF STRATIFIED ROCKS—*Continued*

Periods.	Systems.	Formations.	
PRIMARY, OR PALÆOZOIC.	**PERMIAN** or **DYAS** (1500 ft.)	Red Sandstone, Marl } Zechstein Magnesian Limestone, etc. } Red Sandstone and Conglomerate Rothliegende	Dominant type, Fishes
	CARBONIFEROUS (12,000 ft.	Coal Measures and Millstone Grit Carboniferous Limestone Series	
	DEVONIAN & OLD RED SANDSTONE (5000 ft.)	Upper Old Red Sandstone Devonian Lower Old Red Sandstone	
	SILURIAN (7000 ft.)	Ludlow Series Wenlock Series Llandovery Series May Hill Series	Dominant type, Invertebrates
	ORDOVICIAN (15,000 ft.)	Bala and Caradoc Series Llandeilo Series Llanvirn Series Arenig and Skiddaw Series	
	CAMBRIAN (12,000ft.)	Tremadoc Slates Lingula Flags Menevian Series Harlech and Longmynd Series	
	EOZOIC—ARCHÆAN (50,000 ft.)	Pebidian, Arvonian, & Dimetian Huronian and Laurentian	

The total is 117,500 feet. Estimates in years of geological time, often given in Newspapers, are quite misleading. No geologist can say how many years ago the Eocene, or any other period, began. We only know that over 100,000 feet of rocks have been deposited; and, by taking a rough average (1 foot in 1000 years), we get as the result of multiplying 100 million years. But this is only a very rough estimate. The world may be much older than that, it can hardly be less since geological time began.

APPENDIX II

THE GREAT SEA-SERPENT

MR. HENRY LEE discusses the question of "The Great Sea-Serpent" in an interesting little book, entitled *Sea Monsters Unmasked*, illustrated (1883), published as one of the Handbooks issued in connection with the International Fisheries Exhibition. He goes fully into the history of the subject, and shows how some of the appearances described may be accounted for; but yet is inclined to think that there may exist in the sea animals of great size unknown to science, and concludes as follows :—

"This brings us face to face with the question, 'Is it, then, so impossible that there may exist some great sea creature, or creatures, with which zoologists are hitherto unacquainted, that it is necessary in every case to regard the authors of such narratives as wilfully untruthful or mistaken in their observations, if their descriptions are irreconcilable with something already known?' I, for one, am of the opinion that there is no such impossibility. Calamaries or squids of the ordinary size have, from time immemorial, been amongst the commonest and best known of marine animals in many seas; but only a few years ago any one who expressed his belief in one formidable enough to capsize a boat or pull a man out of one was derided for his credulity, although voyagers had constantly reported that in the Indian seas they were so dreaded that the natives always carried hatchets with them in their canoes, with which to cut off the arms or tentacles of these creatures, if attacked by them. We now know that their existence is no fiction; for individuals have been captured measuring more than fifty feet, and some are reported to have measured eighty feet in total length. As marine snakes some feet in length, and having fin-like tails adapted for swimming, abound over an extensive range, and are frequently met with far at sea, I cannot regard it as impossible that some of these also may attain to an abnormal and colossal development. Dr. Andrew Wilson, who has given much attention to this subject, is of the opinion that 'in this huge development of ordinary forms we discover the true and natural law of the production of the giant serpent of the sea. It goes far at any rate towards accounting for its supposed appearance. I am convinced that whilst naturalists have been searching amongst the vertebrata for a solution of the problem, the great unknown, and therefore unrecognized, Calamaries, by their elongated cylindrical bodies and peculiar mode of swimming, have played the part of the sea-serpent in many a well-authenticated incident. In other cases, such as those mentioned by 'Pontoppidan' (*History of Norway*), the supposed vertical undulations of the snake seen out of water have been the burly bodies of so many porpoises swimming in line—the connecting undulations beneath the surface have been supplied by the imagination. The dorsal fins of basking sharks, as figured by Dr. Andrew Wilson, may have furnished the 'ridge of fins;' an enormous conger is not an impossibility; a giant turtle may have done duty, with its propelling flippers and broad back; or a marine snake of enormous size may really have been seen. But if we accept as accurate the observations recorded (which I certainly do not in all cases, for they are full of errors and mistakes), the difficulty is not entirely met, even by this last admission, for the instances are very few in which an Ophidian proper—a true serpent—is indicated. There has seemed to be wanting an animal having a long snake-like neck, a small head, and a slender body, and propelling itself by paddles.

"The similarity of such an animal to the Plesiosaurus of old was remarkable. That curious compound reptile, which has been compared with 'a snake threaded through the body of a turtle,' is described by Dean Buckland as having 'the head of a lizard, the teeth of a crocodile, a neck of enormous length resembling the body of a serpent, the ribs of a cameleon, and the paddles of a whale.' In the number of its cervical vertebræ (about thirty-three) it surpasses that of the longest-necked bird, the swan.

"The form and probable movements of this ancient Saurian agree so markedly with some of the accounts given of 'the great sea-serpent,' that Mr. Edward Newman advanced the opinion that the closest affinities of the latter would be found to be with the Enaliosaurians, or Marine Lizards, whose fossil remains are so abundant in the Oolite and the Lias. This view has been taken by other writers, and emphatically by Mr. Gosse. Neither he nor Mr. Newman insist that 'the great unknown' must be the Plesiosaurus itself. Mr. Gosse says, 'I should not look for any species, scarcely for any genus, to be perpetuated from the Oolitic period to the present. Admitting the actual continuation of the order Enaliosauria, it would be, I think, quite in conformity with general analogy to find some salient features of several extinct forms.'

"The form and habits of the recently recognized gigantic cuttles account for so many appearances which, without knowledge of them, were inexplicable when Mr. Gosse and Mr. Newman wrote, that I think this theory is not forced upon us. Mr. Gosse well and clearly sums up the evidence as follows: 'Carefully comparing the independent narratives of English witnesses of known character and position, most of them being officers under the Crown, we have a creature possessing the following characteristics: (1) The general form of a serpent; (2) great length, say above sixty feet; (3) head considered to resemble that of a serpent; (4) neck from twelve to sixteen inches in diameter; (5) appendages on the head, neck, or back, resembling a crest or mane (considerable discrepancy in details); (6) colour, dark brown or green, streaked or spotted with white; (7) swims at surface of the water with a rapid or slow movement, the head and neck projected and elevated above the surface; (8) progression steady and uniform, the body straight, but capable of being thrown into convolutions; (9) spouts in the manner of a whale; (10) like a long "nun-buoy."' He concludes with the question, 'To which of the recognized classes of created beings can this huge rover of the ocean be referred?'

"I reply, 'to the Cephalopoda.' There is not one of the above judiciously summarized characteristics that is not supplied by the great Calamary, and its ascertained habits and peculiar mode of locomotion.

"Only a geologist can fully appreciate how enormously the balance of probability is contrary to the supposition that any of the gigantic marine Saurians of the secondary deposits should have continued to live up to the present time. And yet I am bound to say that this does not amount to an impossibility, for the evidence against it is entirely negative. Nor is the conjecture that there may be in existence some congeners of these great reptiles inconsistent with zoological science. Dr. J. E. Gray, late of the British Museum, a strict zoologist, is cited by Mr. Gosse as having long ago expressed his opinion that some undescribed form exists which is intermediate between the tortoises and the serpents." (This is quoted by Mr. Lee in a footnote.)

"Professor Agassiz, too, is adduced by a correspondent of the *Zoologist* (p. 2395), as having said concerning the present existence of the Enaliosaurian type, that 'it would be in precise conformity with analogy that such an animal should exist in the American seas, as he had found numerous instances in which the fossil forms of the old world were represented by living types in the new.'

"On this point, Mr. Newman records in the *Zoologist* (p. 2356), an actual testimony which he considers 'in all respects the most interesting natural history fact of the present century.' He writes—

"'Captain the Hon. George Hope states that when in H.M.S. *Fly*, in the Gulf of California, the sea being perfectly calm, he saw at the bottom a large marine animal with the head and general figure of the alligator, except that the neck was much longer, and that instead of legs the creature had four large flappers, somewhat like those of turtles, the anterior pair being larger than the posterior. The creature was distinctly visible, and all its movements could be observed with ease; it appeared to be pursuing its prey at the bottom of the sea. Its movements were somewhat serpentine, and an appearance of annulations, or ring-like divisions of the body, was distinctly perceptible. Captain Hope made this relation in company, and as a matter of conversation. When I heard it from the gentleman to whom it was narrated, I inquired whether Captain Hope was acquainted with those remarkable fossil animals, Ichthyosauri and Plesiosauri, the supposed forms of which so nearly corresponded with what he describes as having seen alive, and I cannot find that he had heard of them; the alligator being the only animal he mentioned as bearing a partial similarity to the creature in question.'

"Unfortunately, the estimated dimensions of this creature are not given.

"That negative evidence alone is an unsafe basis for argument against the existence of unknown animals, the following illustrations will show:—

"During the deep-sea dredgings of H.M.S. *Lightning*, *Porcupine*, and *Challenger*, many new species of mollusca and others, which had been supposed to have been extinct ever since the Chalk, were brought to light; and by the deep-sea trawlings of the last-mentioned ship there have been brought up from great depths fishes of unknown species, and which could not exist near the surface, owing to the distention and rupture of their air-bladder when removed from the pressure of deep water.

"Mr. Gosse mentions that the ship in which he made the voyage to Jamaica was surrounded in the North Atlantic, for seventeen continuous hours, by a troop of whales of large size, of an undescribed species, which on no other occasion has fallen under scientific observation. Unique specimens of other Cetaceans are also recorded.

"We have evidence, to which attention has been directed by Mr. A. D. Bartlett, that 'even on land there exists at least one of the largest mammals, probably in thousands, of which only one individual has been brought to notice, namely, the hairy-eared, two-horned rhinoceros (*R. Lasiotis*), now in the Zoological Gardens, London. It was captured in 1868, at Chittagong, in India, where for years collectors and naturalists have worked and published lists of the animals met with, and yet no knowledge of this great beast was ever before obtained, nor is there any portion of one in any museum. It remains unique.'

"I have arrived at the following conclusions: 1. That without straining resemblances, or casting a doubt upon narratives not proved to be erroneous, the various appearances of the supposed 'great sea-serpent' may now be nearly all accounted for by the forms and habits of known animals; especially if we admit, as proposed by Dr. Andrew Wilson, that some of them, including the marine snakes, may, like the cuttles, attain to extraordinary size. 2. That to assume that naturalists have perfect cognizance of every existing marine animal of large size, would be quite unwarrantable. It appears to me more than probable that many marine animals, unknown to science, and some of them of gigantic size, may have their ordinary habitat in the sea, and only occasionally come to the surface; and I think it not impossible that amongst them may be marine snakes of greater dimensions than we are aware of, and even a creature having close affinities with the old sea-reptiles whose fossil skeletons tell of their magnitude and abundance in past ages.

"It is most desirable that every supposed appearance of 'the Great Sea-Serpent' shall be faithfully noted and described; and I hope that no truthful observer will be deterred from reporting such an occurrence by fear of the disbelief of naturalists or the ridicule of witlings."

APPENDIX III

LITERATURE

1. Popular Works

Works by Dr. Gideon A. Mantell :—
Medals of Creation.
Wonders of Geology.
Petrifactions and their Teaching.
Phases of Animal Life. By R. Lydekker.
Science for All. 5 vols. (Chapters on Extinct Animals.)
Winners in Life's Race. By Arabella Buckley (Mrs. Fisher).
Records of the Rocks. By W. S. Symmonds.
The Story of the Earth and Man. By Sir Wm. Dawson.
The Old Red Sandstone. By Hugh Miller.
Sketch-Book of Popular Geology. By Hugh Miller.
Early Man in Britain. By Prof. Boyd Dawkins.
Cave-hunting. By Prof. Boyd Dawkins.
The Horse. By Sir W. H. Flower.
The Living World. W. H. Conn.
Extinct Animals. Sir E. Ray Lankester.

2. Works of Reference

A Manual of Palæontology. 2 vols. By Prof. Alleyne Nicholson and R. Lydekker.

The Life-History of the Earth. By Prof. Alleyne Nicholson.

Origin of Species. By Charles Darwin.

The English Encyclopedia. (The 2 vols. on Natural History contain much information on extinct animals.)

The Encyclopedia Britannica. Eighth or Ninth Editions.

Vertebrate Palæontology. By A. Smith Woodward.

Phillips's *Manual of Geology.* New Edition, by Prof. H. G. Seeley and R. Etheridge.

Prehistoric Europe. By Prof. James Geikie.

Palæontological Memoirs. By Hugh Falconer, M.D.

Mammals Living and Extinct. By Sir William Flower and R. Lydekker.

British Fossil Mammals and Birds. By Sir R. Owen.

A History of British Fossil Reptiles. By Sir R. Owen. 4 vols. (Cassell.) Most of this work has been previously published in the *Monographs of the Palæontographical Society.*

A Manual of Palæontology. By Sir R. Owen.

Memoir on the Archæopteryx. By Sir R. Owen.

Extinct Fossil Mammals of Australia. Sir R. Owen. London, 1887.

Odontography. By Sir R. Owen.

A Catalogue of British Fossil Vertebrata. By A. S. Woodward and C. D. Sherborn.

The Geographical and Geological Distribution of Animals. By Prof. A. Heilprin.

Memoirs of the Geological Survey of Great Britain.

Monographs of the Palæontographical Society.

Palæontographica. Edited by Prof. K. A. Zittel.

Text-Book of Palæontology. Vols. I. and II. K. A. Zittel.

Ichnology of New England. By Edward Hitchcock. Boston, 1858.

Ichnographs from the Sandstone of the Connecticut River. By Dr. J. Deane. Boston, 1861.

Ichnology of Annandale. By Sir Wm. Jardine. Edinburgh, 1853.

Fossil Footsteps in the Red Sandstones of Potsville, Pennsylvania. By Prof. J. Lea.

History and External Characters of the Dodo, Solitaire, and other Extinct Brevipennate Birds of Mauritius, Rodriguez, and Bourbon. H. E. Strickland and A. C. Melville. London, 1848.

A Monograph on the Extinct Toothed Birds of North America. By Prof. O. C. Marsh. New Haven, Connecticut, 1880.

The Dinosaurs of North America. Prof. O. C. Marsh.

Also *The Dinocerata.* Prof. O. C. Marsh. Washington, 1896.

The Vertebrata of the Tertiary Formations. By Prof. E. D. Cope.

The Vertebrata of the Cretaceous Formations of the West. By Prof. E. D. Cope.

The Survival of the Fittest. By Prof. E. D. Cope.

Primary Factors of Evolution. Prof. E. D. Cope.

Contributions to the Extinct Vertebrata Fauna of the Western Territories. By Joseph Leidy. Washington, 1873.

(The above Monographs by Marsh, Cope, and Leidy are all in the United States Geological Survey of the Territories.)

Cretaceous Reptiles of the United States. Joseph Leidy. Smithsonian Contrib. Knowledge, vol. xiv. (1864).

Reports of the British Association for the Advancement of Science.

Catalogue, Tertiary Vertebrata of the Fayûm, Egypt. By C. W. Andrews. Trustees of the British Museum.

Papers by Prof. H. F. Osborn in *Memoirs of the American Museum of Natural History, and in Bulletins of Do.*

Fauna der Gaskohle und der Kalkstein der Perm-formation Boehmens. By A. Fritsch. Prague, 1883–1886.

Les Enchainements du Monde Animal. By A. Gaudry. Paris, 1883.

Monograph of the Fishes of the Old Red Sandstone of Britain (*Cephalaspidæ*). *Palæontographical Society*, 1868–1870. J. Powrie and Ray Lankester.

Die Placodermen. C. H. Pander. 1857.

Geological Survey of Ohio. Palæontology, I., II.

3. Journals

The student will find numerous valuable papers by well-known geologists in the following Journals :—

The Philosophical Transactions of the Royal Society.

The Quarterly Journal of the Geological Society.

The Geological Magazine.

Nature.

Knowledge. Papers by R. Lydekker.

The American Journal of Science. Papers by Prof. O. C. Marsh.

The American Naturalist. Papers by Prof. E. D. Cope.

Bulletin of the American Museum of Natural History. New York.

Anales de Museo Publico de Buenos Aires. Papers by H. Burmeister.

Anals de Museo de La Plata.

INDEX

THE END

PRINTED BY WILLIAM CLOWES AND SONS, LIMITED, LONDON AND BECCLES.

EXTINCT MONSTERS.

1892 Edition.

A POPULAR ACCOUNT OF SOME OF THE LARGER FORMS OF ANCIENT ANIMAL LIFE.

With Twenty-four Plates, and numerous other Illustrations: and a Preface by Dr. HENRY WOODWARD, F.R.S.

OPINIONS OF THE PRESS.

"A deeply interesting work."—*Times.*

"If the popularity of extinct monsters has of late not been quite what it once was, it ought to be revived by Mr. Hutchinson's excellent book."—*Guardian.*

"This is undoubtedly the best book that Mr. Hutchinson has written."—*Athenæum.*

"Just such a book as this was urgently needed to stimulate popular interest in palæontology."—*Manchester Guardian.*

"Quite one of the most successful of recent undertakings in the field of popular science."—*Morning Post.*

"A wonderfully interesting volume."—*Review of Reviews.*

"Mr. Hutchinson is a skilful and trustworthy guide through 'the valley of dry bones.'"—*Daily Chronicle.*

"His book, in short, is both attractive and useful, and will add to his reputation as a popular, but accurate, writer on geological subjects."—*Saturday Review.*

"A piece of natural history that is far more amusing than most novels, and as full of instruction as a book of its size can well be."—*National Observer.*

"It is thoroughly readable."—*The Field.*

"The book is altogether beautifully got up, and fulfils, in every sense, that most difficult of requirements—a good and true geological work which is at the same time perfectly popular."—*Journal of Education.*

"This work is an admirable account of some of the larger forms of ancient animal life."—*New York Nation.*

"The work fully sustains his reputation as a popular writer on scientific subjects."—*New York Times.*

"Not before it was needed, the writer of this interesting volume has undertaken a popular and at the same time trustworthy account of the larger form of ancient animal life."—*New York Tribune.*

Printed in Dunstable, United Kingdom

91220129R00268